First Light

FIRST LIGHT

THE SEARCH FOR THE EDGE OF THE UNIVERSE

Richard Preston

Random House

New York

This is a revised edition of a work that was originally
published in hardcover by Atlantic Monthly Press in
1987 and in paperback by Plume in 1988.

Grateful acknowledgment is made for permission to quote
from the following:
    Juan Carrasco's notebooks.
    Maarten Schmidt's notebooks.
    Interview of Maarten Schmidt by Spencer Weart in
October 1977, pp. 100–137, and interview of Maarten
Schmidt by Paul Wright on March 10, 1975, pp. 18–23;
both in the collection of the American Institute of
Physics.
    Interview of James A. Westphal by David DeVorkin on
August 9 and 12 and September 14, 1982, pp. 1–23,
48–53, and 117–168; in the National Air and Space
Museum of the Smithsonian Institution, Space
Astronomy Oral History Project.

ISBN 0-679-44969-8

Random House website address:
http://www.randomhouse.com/
Printed in the United States of America on acid-free paper
Book design by J. K. Lambert
98765432

First Revised Edition

For

Michelle Parham Preston,

my guide star

The desire of the moth for the star,

Of the night for the morrow,

The devotion to something afar

From the sphere of our sorrow.

SHELLEY

Foreword: To Readers and Teachers

First Light is nonfiction, a true story about astronomers who are looking for light coming from the edge of the universe. It tells how science is really done—and science is a lot weirder and more human than most people realize. The book has been out of print and hard to find; this is a revised and updated edition. For some reason, *First Light* has gotten a reputation as a kind of cult classic about science. I never really intended it to be read as a science book, but books, like children, have a way of choosing their own friends.

The action centers on the two-hundred-inch Hale Telescope, or the Big Eye. This telescope is a wonder. It sits inside a dome near the summit of Palomar Mountain, in southern California, not far from San Diego. It was built during the 1930s, and is probably the masterwork of the Depression. It is a huge telescope, the heaviest working telescope on earth. Seven stories tall, the Hale Telescope glides so easily on Flying Horse telescope oil that you can move it by hand. It has a mirror two hundred inches in diameter—sixteen feet, eight inches across. The Hale's mirror took fourteen years to cast and polish. During the final stages, the master opticians polished the glass with their bare thumbs. They made a mirror so smooth that if it were expanded to the size of the United States it would not show any bump more than four inches high.

The telescope was conceived by an astronomer named George Ellery Hale, who suffered from a mental illness that gave him hallucinations. When Hale was having a bad day, a little man, a kind of elf, perched on his shoulder and talked into his ear, giving him all kinds of advice. This elf would not shut up. After a while, Hale felt as though the elf was beginning to drive him crazy. I like to think that the elf whispered something in Hale's ear like: "Hale, you have got to build a telescope to end all telescopes." Whether or not the idea for it came to Hale in this way, his telescope is one of the great achievements of the human spirit. Just to look at it makes one feel a little better about our species. We can do a few things right.

■ ■ ■

The main figure in *First Light* is James E. Gunn, who is regarded by some experts as the leading astronomer of our time. Whatever else he is, Jim Gunn is a genius, pure and simple. He is also a so-called gadgeteer. That is, he builds scientific instruments. Unlike other gadgeteers, he builds his gadgets out of junk parts—Gunn is a cheapskate, but a very clever one. He and his colleagues find some of the parts in dumpsters. On scavenging trips around Los Angeles, the city that was originally Gunn's base of operations (he is now at Princeton University, in New Jersey), Gunn and his cronies discovered dumpsters that were used by military defense contractors. They were mother-lode dumpsters. If you knew what to look for, you might find a fifty-thousand-dollar sensor lying among slices of moldy pizza, drenched with Lipton's Cup-a-Soup. If you cleaned the soup off the device, it might work, or you could at least cannibalize it for parts. Gunn also obtained parts from mail-order catalogs and surplus stores. He got parts from his corner drugstore and from Woolworth's and L. L. Bean. He conned NASA into giving him spaceflight items he couldn't possibly afford to buy. Then he wired and bolted the stuff together into supersensitive instruments, attached them to the Hale Telescope, and used them to see right to the edge of the universe, discovering things no one had seen before. Gunn found light never meant for human eyes. The expression "first light" is a technical term. To "see first light" means to open the eye of a new telescope, allowing starlight to fall on the mirror and the sensors for the first time.

That is when you find out if the thing works. Gunn's stuff doesn't always work the first time. Come to think of it, it *never* works the first time.

▪ ▪ ▪

In the course of writing *First Light,* I climbed all over and through the Hale Telescope, where I found rooms, stairways, tunnels, and abandoned machines leaking oil. My notebooks show tooth-marks where I gripped them with my teeth while climbing around inside the telescope, and the notebooks are stained with Flying Horse telescope oil. At times I had to take notes by starlight, when the moon was down, or in pitch-darkness, writing by sense of touch. I don't use a tape recorder. I write longhand (very fast) in old-fashioned reporter's notebooks. Tape recorders always seem to break down at the worst possible moment, and anyway, a tape recorder can't capture important visual and sensory details of a scene. Eventually I bought an expensive high-tech darkroom flashlight to provide light for taking notes—it had to be very dim, so it wouldn't disturb the astronomers at their work. This one was the dimmest flashlight money could buy. I had to hold it an inch from the paper just to see what I was writing, but the astronomers complained that it "blinded" them, so I had to turn it off. One night, writing notes by touch in total darkness and in bitter cold, I accidentally turned my notebook upside down. Then I started turning the pages in the opposite direction, unaware that I was writing over my notes. Thus I ended up with two sets of notes in that notebook, one set written on top of the other, running in the opposite direction and upside down. I never was able to read that particular notebook.

Eventually the astronomers seemed to forget I was there, and so I became like Jane Goodall among the chimpanzees. I was able to watch them without causing a disturbance, while they fed on Oreo cookies or looked at galaxies on television screens, oblivious to the presence of a reporter scribbling in a corner.

I was surprised to see how chaotic, amusing, and passionate science is. Scientific facts are often described in textbooks as if they just sort of exist, like nickels someone picked up on the street. But science at the cutting edge, conducted by sharp minds probing deep into nature, is not about self-evident facts. It is about mystery and not knowing. It is about taking huge risks. It is about wasting time, getting burned, and failing. It is like trying to crack a monstrous safe that has a complicated, secret lock designed by God. Some of God's safes are harder to open than others. The questions may be so difficult to answer, the safe so hard to crack, that you may spend a lifetime playing the tumblers and finally die with the door still firmly locked. Science is therefore about obsession. Sometimes there is a faint clicking sound, and the door pulls wide open, and you walk in.

▪ ▪ ▪

During the last phase of writing, I put *First Light* through an intense fact-checking process. My wife used to be a fact-checker for *The New Yorker,* and she taught me how to check a manuscript using colored pencils. So I put colored underlining and marks all over the manuscript, showing passages that had been checked or still needed checking. I reinterviewed everyone in the story, running up three thousand dollars in telephone bills. I made constant changes in the text, essentially rewriting the book while grilling my characters over the telephone. Nothing is "made up" in *First Light,* not even the thoughts of astronomers. When I describe the thoughts that are passing through an astronomer's mind, it is because I asked him what he was thinking. And later, of course, I checked it with him, to make sure I had described the flow and structure of his thought in a way that he remembered and was able to recognize.

The astronomers did not always greet my writing style with approval. Maarten Schmidt, an extremely distinguished astronomer—former president of the American Astronomical Society—generally liked the book, but he objected to the way I had described his interest in watching wrestling matches on late-night television. (Scientists don't think their lives are interesting enough to warrant such detail—but I would disagree.) One day Maarten Schmidt, Jim Gunn, and another astronomer, Donald Schneider, were driving in a car to Palomar Mountain, and they got into a discussion about *First Light*'s shortcomings. (I heard about it later.) Schmidt blurted out something like, "I just don't know where Richard Preston got that crazy stuff about my liking to watch TV wrestling! I can't remember saying it to him! He must have made up the quotes! And what is really terrible is the way I seem to demonstrate such detailed knowledge of Hulk Hogan!"

Choking back laughter, his colleagues told him they remembered hearing him speak to me quite knowledgeably about Hulk Hogan. The quotes, they said, were faithful.

At one point while I was drafting the book, I decided to enlarge the childhood biography of Donald Schneider, so I telephoned Don's mother in Nebraska, to get the facts of his childhood. She talked to me for a long time, giving me nice, rich material, as one would expect from a mother. Then she said to me, "I wonder if I could ask you for a small favor. Well, this is a little embarrassing, but I hope you will mention in your book that Don really needs to get married. Some nice young woman could read about him and see what a wonderful person he is."

I thought, Well, why not? Mothers are often right in these matters. So I added a passage describing how Don Schneider was single and looking to get married.

When Don Schneider read *First Light,* he was appalled that I had broadcasted his desire to get married. We had become friends during the writing of the book, but that did not lessen his annoyance. "It is a frightening thing to be in the hands of a writer," he said to me.

Then something strange happened. Don received an admiring letter from a young woman in the Netherlands who had read about him in *First Light.* She had never met Don, but she already knew about him from her brother, an astronomer who knows Don. Don wrote to the young woman. She replied to him. This continued for more than a year, during which time their letters heated up into love letters—it was a romance carried on by mail, as in Victorian times—and one day Don flew to the Netherlands and proposed to her. They are now happily married, and have two children, and are living in Pennsylvania.

Two astronomers in *First Light* ended up being famous, but not because of the book. They are Eugene (Gene) and Carolyn Shoemaker, husband and wife, who were using a tiny telescope at Palomar to search for comets and asteroids that could hit the earth.

These objects are almost invisible. They are as black as coal and as large as Mount Everest, and they come out of nowhere, booming toward the earth from all angles. When one of them hits the earth it practically burns all life off the planet, enveloping the earth in a holocaust that makes a thermonuclear war look like a Sunday cookout. Fortunately this doesn't happen very often. But it does happen: everything changed for the Shoemakers when Carolyn, along with two colleagues, David H. Levy and Philippe Bendjoya, discovered a chain of unusual comets floating through space that looked like "a string of pearls." The comets took a loop around Jupiter and then slammed into the planet, one after the other in a series of violent explosions, during the summer of 1994. Collectively the objects were called Comet Shoemaker-Levy. When they hit Jupiter, the flashes of light were seen by telescopes all over the earth. CNN carried the news live, and the impacts left a string of brown bruises on Jupiter. These impacts were the most spectacular event to occur in the solar system in recorded human history. The Shoemakers became world famous, and they went on a lecturing tour that has not yet ended. As Gene later said to me, "Our lives have disintegrated into chaos. I guess it only happens once, anyway."

And what of James E. Gunn, the gadgeteer? He is still at Princeton University, where he has embarked on one of the most ambitious projects in modern astronomy, the so-called Sloan Digital Sky Survey. The goal of this $40-million project (Gunn's gadgets are not so cheap anymore) is to make an electronic three-dimensional map of the universe, in color, using all the techniques that Gunn pioneered in the story told in *First Light*. Gunn and his group are essentially building a huge color scanner to scan the heavens. The scanner will sit in a telescope on a mountain in New Mexico. If it works, the result will be like having an atlas of the world when before all you had was a sketch map drawn by a monk. The map will contain a million quasars and a hundred million galaxies, and it will show the three-dimensional structure of the creation. It will probably reveal new kinds of objects that nobody suspected.

The thing about nonfiction writing is that a book's characters go on developing after the book is finished. This can be disturbing to the book's author. I have always envied novelists for the way they can maintain control over their characters. If necessary, they can get rid of them by killing them or sending them to Tibet. The nonfiction writer does not enjoy such luxuries. You can't control your characters. Therefore you can't shape the plot. This gives you an unpleasant feeling that your book is out of control. Even so, one of the secrets of nonfiction narrative is its unpredictability, for this gives the book a convincing reality, the fractal surprise of unfolding life. I have always found it difficult to finish writing a nonfiction book, or rather to let the story go, because there always seems to be more to write about, and as you reach the end of the book you begin to perceive that the story will never end. The book must end, but the story flows along like a stream, until it meets other stories of other lives, and they touch and run along together, and merge into the headwaters of history.

—*Richard Preston*
1996

Contents

PART 1

Big Eye

When the alarm clock woke Juan Carrasco, the senior night assistant at Palomar Observatory, daylight was streaming through cracks in the black window shades of the bedroom. He got out of bed and tugged at a shade, which came up with a crackling sound. The shade had seen so much use that it had become crisscrossed with zigzag breaks, which he had patched with a type of transparent tape reinforced with nylon threads and known to the astronomers of Palomar Observatory as Palomar Glue, since it is used by them to fix almost anything that breaks. What he should do, he would say to himself, was get some new tape for these shades. Some black tape. To keep out the daylight, so he could sleep better. He found his glasses and looked over a ridge covered with manzanita to the tops of clouds popping up on the far side of the ridge, like torn cotton: a good sign. A sign of clear skies coming tonight. Juan crossed the bedroom, past a photograph of his wife, Lily, and himself taken on the day that Father Girán had married them, and took a leisurely shower.

Then he shaved. In the mirror, as he pulled foam from his face with a disposable razor, broad cheekbones emerged, under brown eyes. Shaving took a long time. He believed, in fact, that he had never properly gotten the hang of the throwaway razors. He was a former barber. He had learned to be very, very careful with a straight razor when working on a customer, and now he could not help being much too careful with a throwaway razor. He had never cut a customer, not even when one of those winos he used to practice on when he was in barber school slumped over in the barber's chair or began thrashing around. To have a bloody, bellowing customer in the chair would have hurt his pride, and so he had never let his hand slip. An astronomer could groan more abnormally than a sick wino when there was trouble with the Hale Telescope, and so he tried never to let his hand slip at the controls. He rubbed a little grooming cream on his hair, which had begun to silver at the temples, and parted it on the left.

Juan dressed and went outdoors to examine the weather. He stood for a moment in his backyard, before the wild apple trees. Through their bare branches he saw last night's snow on Mount San Jacinto, forty miles to the north, gleaming in the oblique sun. The intervening land was covered with a sheet of fog, but the sky above was creamy yet cloudless, the color of an old blue Chevrolet.

Lily was watching the San Diego evening news in the kitchen. She turned down the volume when Juan came in. He poured himself a mug of coffee while she served dinner, and she asked him who he was working with that night.

Juan Carrasco had a formal way of speaking about his job, the job of night assistant. He said that he was working with Dr. Maarten Schmidt, Donald Schneider, and Professor James E. Gunn. He told Lily that those astronomers had been having trouble with their instruments—a new experiment, something never tried before.

Lily noticed Juan's worry. "Sometimes I wonder," she once remarked to me, "if Juan hates to make mistakes." When Juan had been a young father, he had carried his baby daughters around on pillows—he had been that afraid of breaking them. This man had thought you could break a baby just by handling it. This man had been made for handling the controls of great telescopes.

Juan turned up the television for the weather report. Night fog was coming, with marine winds. That was a good sign, and he began to feel that tonight could turn into a clear night for looking at galaxies. At 5:45 P.M. he fitted his hard hat on his head and picked up his flashlight. "Bueno," he said. "Ya me voy"—"I'm going."

"Que te vaya bien," she said, and kissed him. "That you may go well."

Juan walked along a road that crossed a shoulder of Palomar Mountain, a long hogback ridge, 5,600 feet high, situated in the coastal ranges of southern California, about halfway between Los Angeles and San Diego. Smelling wood smoke mixed with a tang of Pacific Ocean fog, he walked past a grove of fir trees where small stucco houses were almost hidden, belonging to other members of the mountain staff of Palomar Observatory. The road turned through a field of brown ferns and headed for an ivory dome. Deep stands of cedar, white fir, Valparaiso oak, and leafless black oak covered Palomar Mountain, and grassy meadows unfurled among the trees. On dry, sunny slopes grew chokecherry, blue buckthorn, wild lilac, wild coffee with a poisonous bean, and a type of ragged dwarf oak with spiny leaves, called carrasco—Juan's last name. *Palomar* means "dovecote" in Spanish, and indeed, the mountain in autumn and spring fills with shoals of migrating birds. No birds sang on the mountain yet, on this night in early March, because at an altitude of more than five thousand feet, spring came slowly to southern California, but the toads had woken up from their winter's sleep, and in the cold of evening they said *keep, keep,* in voices so halting and tentative that they sounded in pain.

Looking west, Juan saw that the moon had already gone down to its grave. The moonless time of the month had arrived, which the astronomers called dark time. They regarded dark time in spring as the best time for seeing galaxies, because in spring, the Milky Way lay along the horizon, where it would not interfere with the view straight up into the deep. When the Milky Way was high in the sky, it blocked a telescope from seeing into the deep universe. During dark time—moonless nights—in spring, you could point a telescope straight up past the Milky Way into extragalactic space, and there was no moon to wash the blackness from the sky. As Juan rounded a bend and neared the dome, he saw a fog bank hanging over a ridge to the west. He regarded the rising fog as a good omen, as long as it did not cover the mountain. City lights smeared a stain across the sky the color of egg yolk. If the fog socked in the valleys tonight, it would cover the lights of surrounding cities while leaving the skies above the mountain transparent and inky black—perfect for seeing galaxies. The sun had dropped behind the fog, and Juan noted with approval the color of the dying light; it was bluish white—no dust in the air. He knew exactly where the sun was. Exactly. He saw that in about six minutes the sun would set. Palomar Mountain would roll into the terminator of earth-shadow, and a view of the universe would begin to unfold.

The dome looked like Hollywood's idea of a Mayan temple. Juan fitted a key into a tall coffered door, and a small service door opened inward. It closed behind him with a bang that echoed among steel piers. It was dark in there. He flicked on his flashlight and climbed a long flight of stairs. He pushed through a door onto the main floor of the dome, at the base of the two-hundred-inch Hale Telescope. Smelling paint and sweet oil, he touched the brim of his hard hat and looked up. He saw that the shutters of the dome were closed, and that the Hale Telescope was pointed straight up, in its normal resting position. It rose seven stories over his head. The Hale hardly looked like a telescope at all to most people: it was a skeleton tube made of struts and girders. Covered with battleship-gray paint, the Hale Telescope looked more like a terrible weapon than a mirror for making images of time gone by. Even after so many years, one still felt a little bit of fear looking up at that instrument; one felt a little bit of fear, always.

Under the telescope an engineer walked back and forth, wreathed in clouds of vapor, pumping Jim Gunn's camera full of liquid nitrogen, preparing it for the night. Juan opened his locker. His breath steamed in the cold. He pulled out a cardboard box, which filled his arms. He shut the locker and crossed the floor gingerly, mindful of the transparent puddles of oil that bled a little from the telescope most nights. His box read LA VICTORIA MARINATED JALAPEÑOS. He had found it in the trash, and although he had prolonged the box's life by winding bands of clear tape—Palomar Glue—around and around it, the box had grown round and flabby.

The marinated-jalapeños box held Juan's notebooks, which contained arcane lore diagnostic of the Hale's innumerable tics. The Hale Telescope had taken twenty-one years to build, from 1928 to 1949. It contained thousands of components—motors and relays, gears and wheels, pipes and pumps—dating from the 1930s. Parts made by companies now bankrupt or merged. Parts unobtainable. Parts no longer understood. Juan Carrasco considered himself to be a small component in an enterprise that seemed to extend beyond

Palomar Mountain, beyond the United States, beyond, perhaps, the world. He doubted his importance to this enterprise. Although he had spent fifteen years climbing all over that telescope, patting it with a dust mop and crawling through hidden rooms inside the telescope, he felt that the Big Eye remained, in certain ways, a mysterious instrument. He felt that if he and the astronomers were to cease to exist, other people would find a use for the Hale Telescope. "Man is dispensable," Juan liked to say. "Telescopes are not." Feeling a tiny bit of nervousness, he entered a small room tucked into the wall of the dome, called the data room. There he saw Maarten Schmidt. Schmidt was a tall astronomer with curly, silvering hair. Schmidt smiled and said, "Good evening, Juan."

"The valleys are filled with fog, Maarten."

"Ah," Maarten Schmidt said. "Good."

"I say it's going to be clear tonight." Juan crossed the room and spoke to an astronomer who had a beard and glasses. "Professor James E. Gunn," Juan said. "Are we going to see galaxies tonight?"

Gunn grinned and said, "I don't know, Juanito."

There were two other people in the data room. One was a young astronomer with blond hair and a beard, named Donald Schneider. He sat facing a computer terminal, next to a computer programmer from the Jet Propulsion Laboratory in Pasadena named Barbara Zimmerman. She was in her forties, and had brown hair and a broad face, and her hands moved decisively over a computer keyboard. She was hammering out an untested computer program: jazz software. "Hey, Juan," she said without looking up.

"Hello, hello," Juan said.

Juan placed his hard hat and the box of marinated jalapeños on a shelf and sat in a swivel chair. Panels surrounded him on three sides, covered with switches and video screens. He hit a switch, and a set of Vickers pumps began to whine, driving Flying Horse telescope oil onto the Hale Telescope's horseshoe bearing. He checked the temperature of the mirror. It was normal. Tonight the controls of the Big Eye belonged to a man who had once been a barber in San Antonio and Pecos, Texas, himself no astronomer, because nobody in their right mind would let an astronomer touch the controls of one of the most powerful telescopes on earth.

▪ ▪ ▪

Earlier that afternoon, just hours before his team was scheduled to begin a new phase of its search for the edge of the known universe, the distinguished astronomer James E. Gunn was sitting at a workbench in a room called the electronics shop, in a lower level of the dome of the Hale Telescope. Jim Gunn was dabbing with a soldering iron at a small blue metal box. A curl of smoke went up from the box. He blinked, dragged a handkerchief from his pocket, and sneezed. He blew his nose and threw the handkerchief on the workbench. He said, "I seem to have the East Coast bug." He snapped a lid on the box. "I don't know what you'd call this little device," he said. "It doesn't have a name."

The box was a rat's nest of spare parts, the size of a cigarette pack. It contained resistors, capacitors, and a few semiconductor chips, which Gunn had rooted out of bins in the electronics shop. In Gunn's universe such a device is known as a kludge. The word rhymes with *stooge*. The box exhibited one toggle switch. Gunn, who had a way of emphasizing certain words when he talked, said, "This *thing*, whatever you call it, will allow us to take data from the Hale Telescope's camera in a way that's particularly effective for finding quasars. We want to park the telescope and just let the stars go by as the earth turns. That produces a continuous picture of sky, like a long piece of film. Unfortunately, the camera on the telescope was not designed to do this."

Astronomers, for the most part, do not look through telescopes anymore. They look at a television screen, which displays an image of the night sky. Virtually all professional telescopes these days have cameras attached to them, and most of those cameras use electronic sensors. The systems required to operate a modern telescope are similar to the systems used to operate a spy satellite. One needs a giant mirror. One needs an electronic camera that focuses large amounts of faint light onto a small, hypersensitive silicon sensor chip. One needs a knowledge of computer programs and of robots. The difference is that astronomers point their sensors away from the earth.

For the last three days Jim Gunn had been getting one or two hours of sleep a night, which dismayed him, because he felt that he had been sleeping too much, probably because he was running a slight fever. He said, "I can't do twenty-four-hour days anymore. I'm getting too old." His other problem, at the moment, was that he had to deal with a reporter. I was taking notes while Gunn worked.

Gunn was then forty-seven years old and slightly under medium height. He has a beard and heavy eyebrows. He has a bold forehead, a fringe of brown hair going thin on top, and alert brown eyes. He is known and admired all over the earth, the recipient of more awards and prizes than he can keep track of or remember. He is a professor at Princeton University. That evening, he wore a brown cable-knit sweater drilled with one or two moth holes, and greasy blue trousers of the style worn by gas station attendants. The pockets of these trousers had been used to carry objects never meant to be put into pockets, and thus were either ripping out or rotting out along the leg seams. On the floor sat his toolbox. Lettering on it read J. GUNN. It was a Sears Craftsman toolbox. It was crammed with tape, wire, electronic chips, screws, nuts, gimcracks, and many different types of pliers. On the workbench sat a pair of gold-rimmed eyeglasses, which were padded around the nosepiece with a wad of electrical tape. He said the tape helped his glasses sit more comfortably on his nose.

The Hale Telescope, and the three other working telescopes at the Palomar Observatory, are owned by the California Institute of Technology—a place better known as Caltech. Caltech is a small private university located in Pasadena, California. Around the basements and laboratories of Caltech, Jim Gunn and his ilk are sometimes referred to as plumbers. They are also known as the Palomar gadgeteers. They are, in fact, professional astronomers who happen to build their own instruments. Gunn was the plumber for a team of astronomers that had been trying to map the edge of the universe. Such an effort required teamwork, and certainly a plumber. The two other members of the team were Maarten Schmidt, who was the team leader, and Donald Schneider, who was Schmidt's assistant. The team also had the help of various engineers and programmers, including Barbara Zimmerman.

Gunn was beginning to get frustrated with this job. For three years now the team had been looking for quasars of a certain rare type. They had not yet found any quasars of this type. Quasars were points of light that glittered in the depths of the universe; cosmic lighthouses. The team wanted to find and map the locations of a few of the most distant quasars. They believed that in doing so they could trace the contour of an otherwise hidden shore: the outer limit of the optically explorable universe. For this particular attempt the team had been awarded four nights on the Hale Telescope in March, when the moon would be down and the Milky Way would be lying flat on the horizon, affording a view of deep sky.

The team had decided to try to doctor the camera on the telescope so that it would scan across the sky, which would speed up the search for quasars. That would be like panning a video camera across a landscape, except that they would pan a telescope across the universe. The Hale Telescope's camera was an electronic system meant for taking snapshots, not motion pictures. The camera, being full of robotic devices, was controlled by a computer that sat on a floor below the telescope, which needed to be rebuilt for the experiment. The quasar team had hired an engineer named Richard Lucinio to tear down this computer and rewire it. Then, just before the moon waned, Richard Lucinio went to the hospital. In Lucinio's words, "I had something funky in my stomach, and to this day I don't know what it was." Neither did two gastroenterologists, musing over Lucinio's gut and wondering whether to operate. Meanwhile the computer at the Hale Telescope remained inoperable. The quasar team had been given a chance to use the Hale and they did not want to throw that chance away merely because their computer engineer might be dying. Gunn had no choice but to try to save the experiment with a soldering iron.

Gunn lives in Princeton, New Jersey. He took a taxi to the Newark airport before sunrise. He flew to Los Angeles. He rented a car and drove to Caltech, where he picked up the J. GUNN toolbox, and then he headed east and south on Interstate 210. When he arrived at the dome of the Hale Telescope, he went inside the dome and lived there for the next three days, until he had built the kludge.

Gunn turned the kludge over in his hand and contemplated it. It would help the telescope's camera to talk to its computer. "You

don't design a thing like this," he said. "You look and see what
you've got. Then you build it." Gunn figured that he could stick
it like a limpet to the camera. He figured that if he flipped the
toggle switch on the kludge to "scan"—he had written "scan" beside
the switch to remind himself which way to throw it—the kludge
would transform the Hale Telescope into a motion-picture scanner.

He picked up an object that looked like a flare pistol. He said,
"This is an eight-hundred-degree hair dryer." He aimed it at the
back of his hand and pulled the trigger to see if it worked. There was
a whir and a smell of burning hair. "Mm!" he said. "It's working." He
directed the hair dryer at a shrink-wrapped cable emerging from
the kludge, and the cable shriveled. Then, picking up a soldering
iron, a roll of wiring diagrams, and the kludge, Gunn ran out of
the room. Astronomers call a set of nights on a telescope a "run,"
and the term is not a metaphor. Gunn took an elevator up one
level. He stepped out of the doors of the elevator and onto the
main floor of the Hale dome, where stands the largest working
telescope on earth.

The Hale Telescope, which is the size of a small office building,
was bathed in sodium light, and the light revealed gleams and glints
of metal inside the dome. Whether accidentally or on purpose, the
Hale dome is almost exactly the size of the Pantheon in Rome.
Jim Gunn took a moment, as he often did, to run his eyes over
the telescope. He would admit that he could never look up without
a sense of disbelief at the last telescope built by George Ellery
Hale. "More than slightly mind-boggling," Gunn would say of it.
Then Gunn hurried across the floor and ran up a set of stairs into
a steel-mesh cage that hangs from the butt of the telescope. This
cage is directly beneath the telescope's main mirror—a concave
mirror made of flame Pyrex glass two hundred inches (sixteen feet,
eight inches) across and coated with a reflective layer of aluminum
on its upper surface. The mirror has a hole in its center. The
concave face of the mirror looks skyward through the tube of the
telescope and gleams at night with reflections of stars, like a pool
of water resting at the bottom of a well.

Gunn spread his wiring diagrams on the floor of the cage under
the mirror. He picked up a flashlight, which seemed to have died.
"Dammit," he growled, rapping the flashlight on the cage. The

bulb glowed a little. He pointed the flashlight around. The floor of the cage was littered with Allen wrenches, alligator pliers, screwdrivers, and rolls of tape. He pointed the flashlight up at a camera plugged into a socket at the base of the telescope, like a shell loaded into a cannon. "That's 4-shooter," he said. "It's been in place a tad more than a year." 4-shooter is the Hale Telescope's main camera; Jim Gunn built it.

Gunn clamped the flashlight between his knees and aimed it upward, in an attempt to throw some light on his camera, while he reached up with both hands and grabbed a hank of loose cables. He dredged a Swiss officer's knife from his pocket, and with it he sliced open a cable. "Gah!" he grunted. He ripped away plastic tissue. "Come on there!" he said, pulling forth a rattail of multicolored wires. He began to graft various wires that were sticking out of the kludge into 4-shooter's nervous system, using a soldering iron. "On a scale of one to ten," he suddenly remarked, "this crisis is only a twenty." He stabbed the soldering iron at a wire and a puff of smoke went up. "I've seen things go higher than twenty," he added. "There is still a happiness at twenty."

The camera called 4-shooter is a white cylinder, five feet long and two and a half feet in diameter, and it weighs 1,500 pounds. Plugged into the bottom of the Hale Telescope, it pokes through the hole in the center of the mirror. Although the camera is enormous (for a camera), from a distance it looks no bigger than a rivet fastened at the bottom of the Hale. Gunn built this camera from scratch in a Caltech basement known as the Wastebasket, where he received much help and many used parts from certain Caltech engineers and technicians who happen to be experts in the arcana of trash, and who are known collectively as the Wizards of the Wastebasket. In certain ways, 4-shooter resembles a scientific package on a spacecraft: it contains a variety of quartz lenses and mirrors, forests of gold connectors and gold-plated parts, and advanced imaging sensors. In other ways it resembles an outrageous kludge: it contains tangles of stainless-steel plumbing, surplus wires, junk motors purchased at deep discount (for ten cents on the dollar or less), movie projector belts, a broken razor blade, Ensolite foam, piano wire, grease, glue, and small, powdery crystals of dried sweat.

A leading scientific instrument usually remains on the cutting edge of science for a few years, until a better instrument comes along, but the Hale Telescope has been breaking trail into the deep for forty years, principally because of the Palomar gadgeteers. The Hale is no longer the largest telescope on earth—Caltech recently built a larger one, called the Keck Telescope, on Mauna Kea in Hawaii, which has a mirror made of glass segments that add up to the equivalent of a four-hundred-inch telescope. The Hale Telescope is, however, world-class. It contains a suite of hypersensitive instruments, of which 4-shooter is one. These instruments, together with the size of the telescope's mirror, make the Hale one of the best telescopes on earth. The Hale is a masterwork of Depression engineering, the Apollo project of the 1930s. Colossal, welded, gray, aloof, massive, agile, apparently indestructible, and uncompromisingly and magnificently extragalactic, the Hale Telescope stands among all telescopes as the climax of dreadnought design. There will never be another telescope like the Hale, because, in the first place, no amount of money could build the Hale Telescope today, and in the second place, the philosophy of telescope design has changed. A new generation of earth-based telescopes is being built, containing large mirrors hung in airy frames, built more like aircraft than ships. Then there is the Hubble Space Telescope, which is a canister that was tipped into orbit three hundred miles above the earth, from the cargo bay of the space shuttle *Atlantis*. But for the time being, the Hale is the world heavyweight champion. The Hale will likely continue to be regarded as one of the world's great telescopes until well into the twenty-first century.

The Hale is a versatile telescope. In addition to its two-hundred-inch primary mirror, it contains a total of eleven smaller mirrors that can be moved and angled in order to reflect and condense light into various points inside or near the telescope, where instruments can be placed. The Hale is a refinery for light. It collects a huge amount of starlight and pours it into a minuscule area. When 4-shooter is plugged into the Hale, starlight lands on the Hale's primary mirror and bounces up to a small secondary mirror (four feet across) at the top of the telescope. The light then bounces downward into 4-shooter, sitting in the hole in the two-hundred-

THE · TWO · HVNDRED · INCH ~
TELESCOPE · LOOKING · NORTHWEST

The Hale Telescope as drawn by Russell W. Porter in 1939, before the telescope was finished. Even though he could only imagine it in its finished state, Porter captured the grandeur of the Hale. The telescope's tube is the open structure of girders. The prime focus cage, where an astronomer can sit and stare directly into the mirror, is at the upper left of the picture, at the top of the telescope. The mirror is at the bottom of the tube, lower right. The curved horseshoe bearing is toward the upper right of the picture, profiled against the night sky, which is apparent through the open dome slit. (Photograph courtesy of Palomar/Caltech) inch mirror. By the time the starlight enters 4-shooter, it has been narrowed down from a beam two hundred inches across into a beam fourteen inches across. The beam of starlight enters a window in 4-shooter, where it is further shrunk and bounced among mirrors. Finally it lands on four electronic chips known as CCDs. Each chip is the size of a child's fingernail. In the end, the light that falls on the main mirror of the Hale Telescope—209 square feet of starlight in all—is distilled onto four chips having a total surface area equal to one postage stamp.

4-shooter is Jim Gunn's favorite toy. It can take four pictures of the sky simultaneously. These pictures can be joined edge to edge to make a four-paneled mosaic. 4-shooter has taken pictures of newborn stars shining through dust cocoons, and of elderly carbon stars coughing bubbles of hydrogen, and of sheets of gas blown from stars that have exploded and died. The camera has made images of dwarf galaxies, of starburst galaxies, and of elliptical galaxies studded with warty globules of stars. It has unveiled the explosive central nuclei of Seyfert galaxies and has looked into their quasarlike cores, filigreed with sable dust. It has taken snapshots of colliding galaxies flinging away threads of stars as they dance and merge with one another. 4-shooter has exposed supergiant cannibal galaxies feeding upon other galaxies for lunch. 4-shooter has imaged rich swarms of galaxies interorbiting like clouds of gnats, and it has mapped gravitational lenses, which are warps in spacetime that break the light of quasars into double, triple, and quadruple mirages from the dawn of time.

Moving quickly around the cage, Gunn appeared to be a tiny figure fighting a tangle of wires, dwarfed by the immensity of the Hale Telescope. He could not see what he was doing, because he had inherited a set of nearsighted eyes from his father. "I can't see up close, and I can't see far away, either," he would say, explaining that he owned a large collection of Woolworth spectacles of varied magnifying powers that he planted everywhere from Princeton to Caltech to Palomar Mountain in order to have a pair of glasses within reach whenever he needed them, but at the moment he had forgotten to leave a pair of glasses inside the Hale Telescope. He hit a switch.

"Donz," he called into an intercom. He was calling for his fellow astronomer Don Schneider.

"Greetings," a voice crackled.

"I need a pair of young eyes," Gunn said.

A door in the wall of the dome flew open, and Don Schneider ran onto the dome floor and scrambled up the stairs into the cage. In addition to his blond hair and beard, Don had a narrow face and intense, flickering blue eyes. He pulled a wool cap down nervously over his head and said, "It's going to be chaos tonight." He stood well back from Gunn's tangle of wires.

"Take it easy," Gunn said.

"What a disaster," Schneider said. "It looks like it's going to fog up tonight." He informed Gunn that the computer system had recently gone insane.

"Yeah," Gunn said. "That's not a problem. Tell Barbara to write some lines of code to fix it." Gunn held out a fistful of wires. "Will you hold this?" he said. The wires trembled; Gunn had developed the shakes from lack of sleep.

"Have you had something to drink?" Schneider asked, smiling at Gunn's shakes.

"Absolutely not. I have been trying for half a fucking hour to solder *three wires.*"

Their heads bent toward the tangle. They worked fiendishly. Plumes of breath and burning rosin smoked in the cold air. Suddenly a flashbulb went off nearby. Don Schneider glanced around. A group of schoolchildren had arrived, with their teachers, to view the progress of American science. The children stood behind a wall of glass—a viewing gallery for visitors, which runs along one side of the dome. The purpose of the glass is to prevent human bodies from flooding the dome with warm air, which would warp the mirror, throwing the stars out of focus. Warm air would also ripple out through the viewing slit of the dome at night, thereby causing the stars to twinkle. Astronomers hate twinkling stars, because twinkling throws the stars out of focus. Fortunately the glass wall also prevents visitors from hearing profanity, which is a type of noise that can be heard often enough coming from the cage at the butt of the Hale Telescope.

"Sun's going down right on schedule," Schneider remarked.

Gunn laughed an edgy laugh.

They finished splicing wires from the kludge into 4-shooter. Pushing his toe around the floor of the cage, Gunn found a roll of transparent packing tape, the kind that is reinforced with threads of nylon. Palomar Glue. Gunn cut a piece of tape with his knife and taped the kludge firmly to the side of his camera. "Palomar Glue," he said, "is what holds this place together."

▪ ▪ ▪

In the data room next to the telescope, Maarten Schmidt sat hunched over an oak desk, in a pool of light thrown from a lamp. He was the senior astronomer on the experiment. The Principal Investigator. The boss. "You see James in a controlled panic," he explained to me. "That is not unusual." Schmidt was a reserved man, gangly and tall. During his lifetime he had spent around five hundred nights on the Big Eye. He described his role in this experiment as much like that of the manager of a baseball team. His star pitcher—Gunn—appeared to be in trouble. All afternoon Gunn had been rushing around, saying, "Don't worry, Maarten, we're almost ready." Maarten had begun to wonder if he might have to cancel the experiment that night and use the Hale Telescope for some other purpose. That might delay the search for quasars by six months, a year, who could tell? Schmidt had become used to delays. He had been searching for quasars for twenty-two years.

Jim Gunn and Don Schneider walked into the data room. Maarten Schmidt said to them, "I think we had better get to dinner." He added to Gunn, "Are you coming with us, James?"

"Yes, in a minute." Gunn crossed the room and sat down at a computer terminal beside Barbara Zimmerman. She was frantically writing computer code that she hoped would operate Gunn's kludge.

Maarten Schmidt and Don Schneider took the elevator to the ground floor and emerged from the dome into afternoon sunlight. They followed a trail among cedar trees and withered ferns dotted with old snow. They avoided mentioning quasars. Maarten said to Don, "You never saw my mark-zero flashlight, did you? It dated from 1950. It was an Eveready. Now it seems I have lost it." A rooster crowed in the distance.

They descended into a hollow where the Monastery stands, a building where the astronomers visiting Palomar Mountain take their meals and sleep during the day. The Monastery has stucco walls and a gabled roof, and it resembles a summer resort gone a little to seed. Schmidt and Schneider sat down at the single long table in the dining room. Several other astronomers, who were working on other telescopes at the observatory, had already arrived. There was a pile of steaks on the table. Astronomers generally require a massive dinner, because the cold in the unheated domes can grow so bad at night that the only bulwark between the astronomer and hypothermia might be a couple of rib-eye steaks inside the astronomer's belly and a bag of Oreos in his hand. The astronomers talked quietly, over a clink of china.

"We're trying to get 4-shooter to read out at a controllable rate," Don Schneider said.

"At what rate?" asked an astronomer.

"One hundred and forty million bytes per hour," Schneider said.

"That's incredible," the astronomer said.

"We should fill twelve tapes a night with data," Schneider added.

■ ■ ■

At the end of the table sat a woman and a man who listened but did not take much part in the conversation. Carolyn Shoemaker had gray hair, cut in bangs, and brown eyes. She wore a maroon sweatshirt and blue jeans. Her husband, Eugene Shoemaker, had a broad face, salt-and-pepper hair, and a clipped mustache. They were a handsome couple. One would imagine them to be normal grandparents, if one did not know that they spent a good deal of their time roaming the Australian outback looking for giant, eroded craters left by asteroids and comets that had smashed into the earth. Gene said, "We're having all kinds of trouble with our telescope." He was referring to the eighteen-inch Palomar Schmidt Telescope, which stood in a small dome three hundred yards south of the Hale Telescope.

"It's an *old* telescope," Carolyn said with affection.

"One of the guide motors has been stalling on us," Gene said. "I think the motor's brushes are shot." He and Carolyn were on Palomar Mountain to search for asteroids and comets.

Carolyn said, "Gene has to dive under the telescope and start the motor by hand, before the photograph smears."

"I have to move fast," Gene said. "I have to grab it around the driveshaft and give it a spin."

Jim Gunn and Barbara Zimmerman walked in. They sat down and nodded to everyone, and Gunn dragged a steak onto his plate with a fork.

"That's not all of it," Gene Shoemaker went on. "We've got some kind of backlash in the main gear. The telescope is jumping all over the place. We can't hold it on a star."

Jim Gunn said, "It sounds like the gears are worn, Gene."

"Exactly," Gene said.

"There's the problem," Jim said. "The gears need a weight on them."

"Exactly," Gene said.

"Get some rope and a two-by-four," Jim said. "Lash the two-by-four to the telescope. Then hang a piece of *lead* on it."

Everyone laughed, including Gene Shoemaker. He saw that Gunn actually had a point there, and he reminded himself to bring some pieces of lead with him the next time he and Carolyn visited Palomar Mountain.

▪ ▪ ▪

There is a saying among astronomers that five billion people concern themselves with the surface of the earth, and ten thousand with everything else. These people are the practitioners of what is said to be the world's oldest science. The astronomers conduct their craft from the vantage point of a droplet of iron and silicates orbiting a G2 star that is now drifting at the inner edge of the Orion Arm of the Milky Way. The Milky Way is a spiral galaxy containing approximately one hundred billion suns. If the Milky Way has other names, the astronomers do not know them yet. They have made *some* progress in the twentieth century, having learned that the Milky Way is a member of what they call the Local Group. The Local Group is a clump of several dozen galaxies, including the Andromeda galaxy and the Clouds of Magellan, that together constitute a virtually unnoticeable knot of galaxies near the outskirts of the Local Supercluster, which is a cloud of many thousands of galaxies. If a galaxy were a leaf, then a supercluster would be the size of a tree. The Local Supercluster amounts to about one-millionth of the observable universe, which throngs with superclusters in the way that a forest is populated with trees. In the more distant parts of the astronomers' universe—as they see it—the brilliant lights called quasars gleam with a physical power that transcends any forces the astronomers have noticed anywhere near the earth. The astronomers do not fully understand quasars— what they are or how they burn—although many quasars are bright enough to be seen with a modest amateur telescope.

Maarten Schmidt hardly touched his dinner. He seemed preoccupied. There was a smell of coffee in the air.

Barbara Zimmerman said to Maarten Schmidt, "I think we've gotten Jim's little box going."

Maarten tapped his fingers on the table and turned to Jim. "Well, James, what next?"

"Keep working, I guess," Gunn said. He and Zimmerman suddenly stood up and walked out of the room.

Schmidt laughed. "I didn't mean that literally!" he called after them.

▪ ▪ ▪

At twilight, on a catwalk that encircled the Hale dome, Juan Carrasco and Don Schneider studied the rising fog, which seemed to be pooling in nearby valleys. Don craned his neck, looking up. He said, "What do you think the weather's going to do, Juan?"

Juan pointed to the west. He said, "There's Venus."

"I suppose you consider that a good sign."

"Oh, yes," Juan said.

"This fog is definitely getting worse," Don said, glancing around.

Juan studied a gauge. "The humidity isn't bad."

"That's if you believe instruments."

Juan slapped the wall, feeling for dew. "Tricky," he said.

"There's definitely some structure up there in the atmosphere. Kind of scuzzy," Don said.

"High haze can improve the seeing."

"Have you ever thought of running for political office, Juan?"

They walked around to the north side of the dome, traveling clockwise on the catwalk. The fog had drowned the Los Angeles basin but had left the San Gabriel Mountains bare and stark on the horizon. The peeping toads had grown louder, welcoming the fog. There were three other telescopes in operation on Palomar Mountain, apart from the Hale Telescope, and as Juan and Don circled, they could see the dome of each telescope in turn: the forty-eight-inch Schmidt telescope, which was being used to make an atlas of the sky; the eighteen-inch Schmidt telescope, which these days was used mostly by the Shoemakers and other planetary astronomers to search for asteroids that could hit the earth; and the sixty-inch Oscar Mayer Telescope, a general-purpose instrument endowed by the family of the hot-dog baron, because Oscar Mayer had liked stars.

Juan made up his mind about the fog. "I think we'll be all right," he said. Don nodded, went indoors to a red button located on the inside wall of the dome, and pressed it. The dome shutters began to move open, drawing apart like eyelids, exposing the telescope to the cosmos. A knife-blade of sky appeared overhead, slowly fattening into a crescent speckled with early stars. The Hale Telescope became a network of shadows against the sky.

The job of the night assistant is to operate the telescope for the astronomer. This not only promotes efficiency but also prevents the astronomer from wrecking the telescope. Given half a chance, an astronomer will cleverly destroy a telescope. For that reason the Palomar night assistants had been given authority over the astronomers in many matters, especially when it came to deciding whether to open or close the dome of the Hale. This was an important decision. For example, a professional astronomer, famished for light, might open the Hale during cold, humid weather. That could let a dew settle on the mirror. The dew could mix with dust on the mirror, which would turn into an acid mud that could etch the glass, thus destroying the mirror in a few hours. The Hale mirror is the size of a living room floor. It weighs fourteen and a half tons. It required fifteen years to make, from the first failed casting in 1934 to the final figuring of the glass in 1949—ground down to a concave dish and polished to a precision of four-millionths of an inch over its entire surface. Four-millionths of an inch is a distance equivalent to splitting the thickness of this page one thousand times.

Inside the data room, Jim Gunn typed instructions to the computer. Maarten Schmidt sat at his desk. Schmidt said, "I cannot tell if they are going to get everything fixed. I am not as teh-nically developed as Jim." He pronounced the word *technically* with a soft *ch*—a Dutch accent. The recent unaccountable loss of his 1950-model Eveready flashlight had touched Maarten Schmidt with a small sorrow, a fact that revealed the nature of his feelings toward electronic devices. But he knew what he wanted from the sky—quasars—and who could help him get them—Jim Gunn. Maarten leaned back and put one ankle across the other knee; he was six feet, four inches tall and did not fit well under desks. He said, "This run is radically new. We have been looking for quasars at exceedingly high redshifts. Tonight is the first time we have ever tried to do this with 4-shooter. It is the most difficult thing we have tried so far—"

"If you do that, Jim"—Barbara Zimmerman's voice rose above the others—"it clears the register."

"Should I go for it?" Gunn asked.

"I don't know," she said. "Hell, yes."

A clatter of computer keys, and then, "Oh, Lord, now what?"

After a slight pause, Maarten Schmidt continued. "The statistical material on these high-redshift quasars is small. We don't know much about their properties. We barely, if at all, understand quasars. Their fate, while they live, is purely speculative." He stood up and squared his stack of papers. He crossed and recrossed the small room while he talked. He said that quasars whose light was shifted strongly toward the red end of the color spectrum were the most distant objects that an optical telescope could resolve—some of them lived, so to speak, at the edge of the universe. Quasars were not easy to find. In all of his team's previous searches for deeply redshifted quasars, they had not found any of this class of object at all. "This is puzzling," he said. "We know they are out there. So why aren't we finding any? No, I am not worried, it is probably a matter of statistics. Perhaps there are not so many of these high-redshift quasars as we had originally supposed. But when you don't find a thing and you know it is there, then there is always the worry that you are doing something wrong."

Quasars are the most luminous objects in the universe. Although they shine at great distances from the earth—far, far beyond the Milky Way—they are so intrinsically brilliant that they appear in a telescope as points of light, like stars. They are not stars. Quasars are tiny objects, however. The core of a quasar may be no larger than a solar system. The energy that causes a quasar to shine is mysterious. Quasars do not "burn" in either a chemical or a nuclear sense of the word. Whatever energy powers a quasar, it is not the thermonuclear fusion that makes the sun shine.

Most astronomers believe that quasars are a long distance away from the Local Supercluster—a long way from our neck of the woods. According to Hubble's law, which is named after its discoverer, Edwin Hubble, the galaxies are moving away from one another. The universe, as an object, is in a state of expansion. Since the totality of human civilization occupies a stroboscopic instant in the unraveling of cosmic time, objects in the sky appear to be motionless, as if caught in a strobe flash, when, in fact, a dance is happening out there. Some galaxies are spinning on their axes, and some galaxies are circling around each other. Two galaxies can touch for a while in a pas de deux, or one galaxy can burst through another, tearing both apart. At the same time, in general, galaxies are slowly withdrawing from one another, because the universe is expanding. Astronomers can discern such movement only through measurement. Spectroscopy—the division of light into its component wavelengths—reveals not only that spiral galaxies are spinning, but also that our galaxy is receding from virtually all other galaxies; that all galaxies are receding from one another (except those bound into clusters by mutual gravity). The galaxies are scattering, somewhat like a crowd leaving a stadium. This general expansion of the universe is called the Hubble flow. The galaxies are being carried along in the Hubble flow. As a result, the light of most galaxies—as seen from earth—is stretched downward in frequency toward the lower end of the color spectrum, toward red. This is a phenomenon known as a Doppler shift. It is similar to the lengthening whistle of a departing train. According to Hubble's law, the more distant a galaxy is, the faster it is moving away from us, and therefore the more downwardly stretched—more redshifted—the galaxy's light appears to be. Quasars are the most redshifted objects in the universe, which to most astronomers signifies that quasars are also the most distant objects that can be seen through a telescope. They inhabit the outer reaches of the optically explored universe: the edge of the universe.

The word *redshifted* is misleading in the case of quasars, for a deeply redshifted quasar is not exactly red in color. Quasars disgorge opulent, multitudinous colors all at once—gamma rays, X rays, ultraviolets, blues, greens, yellows, reds, infrareds, microwaves, and, in the case of some quasars, radio waves, all of which are forms of light at different wavelengths. The trick of recognizing a redshift in a quasar by examining its light was at one time not easy to accomplish. Maarten Schmidt invented that trick in 1963. While reading the text of light from a quasar, Schmidt discovered that quasars are not nearby stars, as everyone had supposed, but monsters—objects on the backdrop of the sky, unimaginably far beyond the local galaxies. In effect, he showed that what looked like fireflies in our backyard were beacons near the horizon.

As a telescope looks out at quasars it looks not only toward the edge of the universe, but also toward the beginning of time. Light consists of photons, which are inseparably both waves and particles. Light moves at a speed of 186,282 miles per second through space—a snail's pace by the measure of cosmic distance. Nothing can move faster than a photon. A photon would require about fifty thousand years to traverse the length of the Milky Way galaxy. If a star were to explode on the other side of the Milky Way, astronomers would learn about it fifty thousand years later.

An event cannot be seen, or known, until the photons emitted by that event reach a detector, such as photographic film or the retina of the human eye. A light-year is the distance that a photon can travel through a void in one year, which happens to be about six trillion miles. Photons produced by an event happening billions of light-years away from an observer will require billions of years to stream toward the observer. When a telescope makes a photograph of deep sky, it makes an image of the past; it displays events that took place in different periods of cosmic history, depending on how far away from the earth they are.

Astronomers refer to the depth of astronomical vision as lookback time. Seeing outward is equivalent to looking backward in time, because the telescope's mirror is capturing primeval light. The universe—as we see it—could be imagined as a series of concentric shells centered on the earth—shells of lookback time. The shells closest to the earth contain images of galaxies near us in time and space. Farther out are shells containing images of remote galaxies—galaxies as they existed before our time. Still farther out is the shell of the early universe. Some of the photons reaching a telescope's mirror are nearly as old as the universe itself. The quasars are brilliant pinpoints of light that seem to surround the earth on all sides, shining out of deep time. Beyond the quasars, the observable universe has a horizon, which could be imagined as the inner wall of a shell. This horizon is the limit of lookback time, which is also an image of the beginning. As a mirror looks toward the edge, it looks toward the beginning. At the end of the sky lies the beginning.

The sky could be imagined as a palimpsest containing stories written on top of one another going back to the origin of time. A telescope looking outward into lookback time strips layers from the palimpsest; it magnifies and reimages small, faint letters in the underlayers of the manuscript. The sky could also be imagined as a book, bound into chapters that tell a story. As a telescope probes out into the sky, it reads backward through the story, from the last chapter to the first. When a mirror collects the light of a distant quasar, it collects photons that have streamed freely through space for most of the time the universe has existed and are now reaching the mirror. The light of a deeply redshifted quasar is a light coming out of chapter one—from somewhere in the middle of the book of Genesis.

The light of the most remote quasars left them at a time when the universe was about ten percent of its age today, evolving rapidly and violently and probably still organizing itself into galaxies. The exact time of this epoch is uncertain, because the age of the universe is uncertain. The universe is probably somewhere between ten and twenty billion years old, which means that the earliest epoch of the quasars happened between nine and eighteen billion years ago. While the quasar's light was traveling on its way to the earth, the quasar died out. A high-redshift quasar is a fossil image—its light exists as a trace of an extinct object. Quasars once shone (are seen shining, will be seen shining) within cosmic history—long before the sun, the earth, and perhaps even the Milky Way came into being, when the universe was young and quite obviously different from today.

The problem of mapping the sky's structure troubles modern astronomers. A photograph taken with a powerful telescope projects the sky onto a two-dimensional surface littered with dots. Some might be asteroids. Many are stars. Even more are galaxies. Some might be comets. A few are quasars. Quasars resemble faint bluish or yellowish stars in a photograph. In appearance, quasars too nearly resemble stars to be winnowed easily from clouds of foreground stars within the Milky Way. For much of his professional career, working step by step outward into space and backward into time, Maarten Schmidt had sought to make a map of quasars in depth—to understand their evolution through time as a class of objects. He had confronted a series of related questions. When were the quasars born? When did they die out, as a species? How did their brightness and their population change while they lived? He wanted to know how the quasars had fared en masse. He wanted to understand the birth, life, and death of the quasars over the range of cosmic time; he wanted to know the natural history of the species.

Quasars seem to be exclusively distant objects. They are rare enough in our neighborhood so that the Local Supercluster, for example, does not contain any quasars. Nearby superclusters do not contain any quasars, either, but if one looks outward through perhaps twenty or thirty superclusters, one begins to notice quasars. The farther out in space (or back in time) a telescope probes, the more quasars it finds. This implies that the quasars gleamed and then gradually died out—and are now dark or dim objects. If quasars still existed today they would be scattered among nearby galaxies. To focus quasar light with a mirror is to reimage the past, since the only optical trace of the quasars today is a memory transported in antique light. The region of quasars begins about two billion light-years away from us, or rather, before our time, and where it all ends, or rather, begins, is what Maarten Schmidt was trying to find out.

Schmidt and other astronomers had noticed that as they looked deeper into the universe and back into the past, the number of distant quasars rose steeply for a while and then seemed to drop off, as if the realm of quasars had an outer surface. At extreme distances, quasars were rare. Astronomers looked through a veil of quasars into apparent darkness. They had reached what seemed to be the edge of the optically known universe. They had touched a kind of membrane beyond which they were unable to see anything in any wavelength of light, apart from the radio hiss of the Big Bang. The astronomers had arrived at a dark time. Beyond the realm of the quasars stretched the visible darkness of the early universe, out of which came no detectable light. Astronomers called this the redshift cutoff. It was the horizon of the quasars. The early universe, before the age of quasars, seemed to be a dim shell surrounding the quasars, beyond the redshift cutoff. The cutoff corresponded to the time at which quasars had first appeared. How and when the quasars had popped on remained a mystery. Quasars seemed to have appeared without precursors or any kind of warning signal. "I suspect that the rise of quasars may have signaled the birth of galaxies," Maarten said. The earliest quasars seemed to be associated with cataclysmic events that had occurred as clouds of hydrogen that had filled the early universe had condensed into galaxies full of stars—a time when the universe had been compact, turbulent, undergoing strong, rapid evolution. Maarten Schmidt felt that if his team could map the rise of quasars over time, they might get a glimpse of the architecture of the creation.

The theorists babbled on at conferences about the nature of the early universe. Maarten Schmidt found something amusing in all this talk. "The theoreticians"—he smiled—"the theoreticians are so clever. Once we have found something, they can find four ways to explain it." In the case of quasars, the theoreticians had already found at least four ways to explain the redshift cutoff *before* it had been mapped. "In all these discussions," Maarten said, "you find that you need hard numbers. How many quasars of this redshift? How many of that redshift? How many quasars of a particular luminosity? It seemed that the bullet had to be bitten."

Exploring the edge of the universe was a dirty job, but somebody had to do it. Somebody had to look. Somebody had to commit quite a few years of a scientific career toward the goal of mapping the realm of the earliest quasars—a gamble that so far had not shown signs of a big payoff. Schmidt and his people had already fed plenty of nickels into the slot machine. Schmidt guessed that years of work lay ahead of him before he might see the structure of the redshift cutoff—if ever. Astronomers learned not to count on results. "This exercise," he said, "is to gather in the facts. You cannot, of course, solve all the questions in the field. Sometimes you get answers to questions you didn't ask."

Edwin Hubble had showed that the redshift of a galaxy depended on its distance from earth: the more redshifted the galaxy, the farther away it was. Much to their regret, astronomers had not yet been able to link this redshift scale to an absolute measure of distance. Thus they could not tell precisely how far away from earth any galaxy or quasar was, except within a relative range. But even if he could not know exactly how far away the quasars were, Maarten Schmidt felt that if he could collect a sample of quasars of the highest redshifts and plot their redshifts on a graph, then he could learn things about the birth of quasars. He wanted to see the distribution of quasars through time in the vicinity of the redshift cutoff.

He had urged Gunn to try to jury-rig 4-shooter, so that the camera would scan long ribbons of sky. To gather a tapestry of stars and then to search the tapestry for quasars might be a good way to find a lot of quasars. The strips of sky would be recorded on magnetic tape. When enough tapes had piled up, from a variety of scans, then Don Schneider, the team's image-processing expert, would analyze the tapes, searching for stabs of quasar light. They could sweep the Hale Telescope across the sky, of course, but it seemed easier to stop the motion of the telescope and let the turning of the earth move the sky past the mouth of the telescope. This was a technique known as a transit. To put a telescope in transit, you shut down the telescope's drive motors, clamp the bearings, and let stars drift past the mouth of the telescope as the earth turns. Instead of scanning the telescope across the sky, Maarten thought that he would let the earth do the work.

The telephone rang in the data room.

Don Schneider picked up the receiver. "Big Eye," he said, and then, after a pause, "It's going to be just unbelievable. Amateur night at the Big Eye—"

"Don't say that, Donz!" Gunn barked.

Don stretched the cord to get out of Gunn's hearing. He lowered his voice and said, "We've been working for three days and we still haven't been able to get the experiment to work. We were just lucky it snowed last night—we couldn't have started, anyway."

Maarten Schmidt, the Principal Investigator, crossed the room and looked over Jim Gunn's shoulder while Jim clacked at the keyboard of the computer.

Jim muttered, "I don't know what will happen tonight, Maarten."

"That's an exciting way to start," Maarten said pleasantly.

"Juan, we need to slew," Jim said. He wanted to move the telescope rapidly (slew it), to point it at a bright star in order to calibrate the sensors.

The night assistant hit a switch. Nothing happened.

Maarten said, "Is there a problem, Juan?"

"No," Juan said, running out of the room.

Maarten laughed. "The telescope has rusted!"

Juan returned. Somebody had left the stairs under the telescope—the telescope always refused to move until the stairs were rolled away. Hitting toggle switches, Juan slewed the telescope across the sky and centered it on a bright star. Jim Gunn typed a command to 4-shooter on his computer keyboard: EXPOSE.

4-shooter responded on the computer screen: OK. But nothing happened.

Gunn peered into the main video screen, which displayed whatever the camera saw. "Black!" he said. "I don't see any stars!" Everyone talked at once.

"4-shooter is real unhappy."

"Something's getting in the way."

"Maybe we're pointed at the ceiling."

There was a pause, and then, "Naw, that's not the problem!"

"Well, where is the dome pointed?"

"We're looking east."

"Is the mirror open?"

"The mirror is open."

"Yeah, but I don't see any stars!" Gunn groaned. "Where's my calculator?" He tore through a pile of papers.

Maarten paced back and forth. He began to whistle "The March of the Wooden Soldiers," from *The Nutcracker*—whistling in the dark, so to speak.

Juan said, "The mirror is open. The dome is open. The stars are out—"

"We're getting no light!" complained Jim Gunn.

"No evidence of light?" Maarten asked.

"Absolutely none at all."

By now they understood that 4-shooter had gone dead.

■ ■ ■

An hour later, the telephone rang in the data room. It was Jim Gunn's wife, Jill Knapp, calling from New Jersey. She is a radio astronomer.

"Hello, love," Jim said, and after a pause, "We're getting light, anyway."

Maarten Schmidt laughed, and to the others he said, "It's like Galileo all over again. First you have to get light down the tube."

Jim and Jill talked quietly. She asked him how he was feeling. He said he felt all right. She asked him if he was getting any sleep. Of course not, he said. Jill Knapp, who is Scottish, once described the frantic hours before an observing run this way: "There is a feeling amongst us astronomers that time on a telescope is extremely precious. One feels the privilege. What a peculiar idea it is for a species to be looking at those . . . " She paused, seeking a word. "At those *things* out there."

The team settled into more tinkering. Jim Gunn kept sending Don Schneider on errands up to the telescope, to flip the switch on the kludge one way or the other. Meanwhile Barbara Zimmerman finished writing her jazz program and left for Pasadena—she had done all she could. Examining a few white blobs on the television screen—stars—Gunn said, "Let's get this pig focused." The night had grown slightly colder, shrinking the Hale Telescope by one-half a millimeter, throwing the stars out of focus. Hitting switches, Juan Carrasco said, "You have the focus." Juan handed a box with buttons on it—a control paddle—to Don. Don tweaked buttons on the paddle, which moved the secondary mirror at the top of the Big Eye by a few hundredths of an inch, and a motor pinged, echoing in the dome. They tested different focuses.

"Ten north," Jim said to Juan.

"Ready," answered the night assistant.

"Ten north again."

"Ready here."

"Twenty north."

"You have it."

The stars on the screen eventually grew sharp. "Delicious!" Juan said.

"Focus disabled?" Don asked Juan.

"Focus off," Juan said.

Jim Gunn turned to Maarten Schmidt. He said, "The telescope is yours."

"What are we going to do?" cried Maarten.

"We can try a transit."

"Really!" Maarten turned to Juan. "Let's shut down the pumps, lights, everything."

Juan said, "Pumps off." The oil pumps shut down, the dome became silent, and the telescope settled and locked onto its bearings. "Lights off," Juan said. The dome went black inside.

"Tracking off?" asked Maarten.

"That is correct," said Juan.

"Somebody has to hit the switch on Jim's box," Don reminded the others. Juan ran out of the data room, followed by Maarten. Juan rolled the stepstairs under the telescope, and Maarten loped up the ladder into the cage, his flashlight bobbing. Overhead stretched a hanging cloth of stars—the dome slit, open to the north. Maarten flipped the toggle switch on the kludge to "scan." He climbed down; Juan rolled the stairs away. In the data room, Jim Gunn punched a computer key and the video screen went dark. A moment later it filled with galaxies. The galaxies were streaming upward across the screen.

"Whoo!" said Schmidt, breathing hard, slapping a pencil against his palm.

"What are we seeing there?" Don Schneider yelled.

"There's a galaxy," Juan said.

"Look at that star truck by," Don said.

"James! By God!" Maarten said. "Wow! Look at that! Aw, fantastic! We are doing it!"

Gunn smiled. It was a small Gunn grin.

They watched the sky move past for a while, and then they decided to shut down the system. It had been a test. "You can stop the sky now," Schmidt said.

Gunn hit a key and the galaxies stopped moving.

"I felt a little jerk," Maarten said, and everyone laughed. He added, "You wouldn't believe the environmental impact statements we had to file to get the sky to do that."

The astronomers decided to try it again. Gunn hit a key, and galaxies began climbing upward across the video screen. A look of disbelief spread over Gunn's face. He could not believe that the kludge was working.

The apparent motion of the night sky, caused by the earth's rotation, rapidly drew objects up the screen, because Hale Telescope's field of view covers only a tiny piece of sky. The video monitor displayed the sky's motion as if the sky were moving from the bottom of the screen to the top: a cloud of galaxies would erupt at the bottom of the screen. In a short while the galaxies would reach the top and disappear one by one, like small bubbles floating up a glass of beer. A computer downstairs was receiving the view and writing it onto spinning tapes.

"How serious is this exercise?" Maarten wondered. Too bad he asked—the screen went blank. 4-shooter had bombed again. Gunn and Schneider pounded keys, while Schmidt paced the room and whistled. Half an hour later, 4-shooter came back to life, and galaxies again floated up the screen. The telescope was looking north, into rotating sky just below the bowl of the Big Dipper. Faint galaxies spattered the screen—usually twenty or thirty galaxies at once—puffs of lint, pinwheels, pearls, eggs, coins. Some galaxies were close and large, but most appeared tiny, resembling bits of confectioner's sugar sprinkled on black velvet. Occasionally a foreground star inside the Milky Way would splash a brilliant, lingering track on the screen as it passed. The view gave a palpable sense of depth to the sky. A big telescope pointed in any direction except into the plane of the Milky Way sees many more galaxies than foreground stars. If you took a hamburger bun and picked one of those little black seeds off it—a poppy seed—and you held the seed out at arm's length, it would just cover the Hale Telescope's field of view when using 4-shooter. In certain areas of the sky that small, 4-shooter had taken snapshots that held as many as two thousand galaxies. This suggests that the sky is largely unexplored. The universe is rather well explored (by human astronomers) out to a distance of about two hundred million light-years from the Milky Way. Farther out, things become vague, while the edge of the knowable universe has been charted with somewhat less conviction than that of the monk who drew the Americas as a little island west of Greenland.

A pair of wretched galaxies with twisted arms passed. They appeared to be puffed up and ensnared in each other, throwing away tendrils of stars.

"Gaw!" said Maarten. "Ejecting galaxies."

A while later, a line crossed the screen.

"What was that?" asked Juan Carrasco.

"Somebody flew by," said Don Schneider.

"It was a meteor," said Jim Gunn.

The Principal Investigator pulled out a pocket calculator. He examined the width of the line and the angle at which the line had traversed the screen. He said that this was not a meteor. He punched some buttons on the calculator. He said that the thing appeared to be a fat, blurry object on a low orbit that passed near the poles of the earth, which would make it a military satellite, probably a spy satellite.

Don remarked that somebody must be watching. He dug up a bag of Chips Ahoy! cookies and passed them around. He settled into a chair with a fistful of cookies, for a night of television. "Man," he said, munching a cookie, "this is pretty good stuff." There were several viewing screens in the data room, including a main screen, which was a video box that sat on a table. All four of 4-shooter's cameras were gathering strips of sky, but the screens showed an identical view, a scene from one of the four strips.

Maarten tapped the main video screen with a pencil. He said, "That's a galaxy, that's a galaxy, that's a galaxy, that's a star. These objects are mostly galaxies." Some of the fainter objects, he said, would be giant elliptical galaxies near the limits of the Hale Telescope's ability to collect light. They were fossil images—their light had started toward the Milky Way more than five billion years ago, before the earth had condensed. The telephone rang. Maarten picked it up. "Big Eye," he said. "O, hallo." It was his wife, Corrie, calling. They chatted quietly in Dutch. He said, "Ja? . . . Dat het gaat regenen? O." He turned to the other astronomers. "She says there's a pressure area forming in the Pacific. We should be all right for another day or two, then we could get some rain."

Jim Gunn took off his glasses and knuckled his eyes and then stared at the screen in silence. As far as Gunn was concerned, he would have to be taken out of the dome in an ambulance before he would give up a night of watching galaxies. Every hour or so, Don Schneider went downstairs and removed a tapeful of sky from the computer. Over time, a cardboard box in the data room filled with tapes.

Juan Carrasco watched galaxies slide up a video monitor in his control panel. He leaned over to speak to the Principal Investigator. He said, "You have a good night on your hands, Maarten."

Maarten grinned.

Don turned to Juan. "Do you like these kinds of nights, Juan?"

"Oh, yes. Nothing to do but look."

"The days are worse than the nights," Don said. "You should have been here this afternoon, Juan. Haven't you noticed my receding hairline?" He pulled a fistful of hair back from his forehead, and Juan laughed. "These have been days of infamy," Don added. Something caught his attention on the video screen. "Here's an unhappy galaxy," he said.

Maarten took off his glasses and squinted at it. "Ja. It's strange-looking, Don, I agree."

"It's giving birth."

"Ha! To a quasar!"

Later a shotgun-blast of galaxies crossed the screen. "Whoa," Don said. "A big, rangy cluster. Some of these rich clusters have a thousand galaxies in them."

"This movie really should be in color," said Maarten.

"Uh-uh, black and white," Don said. "I'm a conservative. Look at that one, Maarten. That's a poor excuse for a galaxy."

"It really is. And what is this object?" Maarten said, pointing to a pair of hazy dots.

"I suspect that's a galaxy with a friend," Don said. The telephone rang again. Don kicked the floor with a pair of black sneakers, and his chair rolled backward across the room to the telephone. "Big Eye," he said into the receiver. "Hello, John! Yes! We are transiting! I know. I know—I was skeptical, I would be the first to admit it."

Schmidt tapped the television screen with his cookie. A huge crowd of galaxies had appeared. He leaned over to speak to the night assistant. He said, "Now there you go, Juan. Three big ovals— a spiral; three more ovals—an irregular." More galaxies appeared. "Look here, we're not done yet!"

Juan pointed to a galaxy. "There's a *very* strange spiral, Maarten."

"Ja, a barred spiral. Nice." Maarten stood up and walked around the data room for a while. Suddenly he wheeled toward the screen and pointed at two disk galaxies locked in an embrace. "Ho! These are in collision!"

"Can you hear these people?" Don raised his voice into the telephone. "We have two interacting galaxies going by." 4-shooter had imaged a pair of galaxies in flagrante delicto, spinning through each other and spraying off stars. The ecstasy would last for a hundred million years. Don finished his conversation, and then the astronomers watched in silence, until perhaps the thought of these fields of galaxies buried in lookback time drew the conversation to time travel. Don said to Maarten, "Did you ever see that *Star Trek*, the one where Captain Kirk ends up back in New York in the 1930s?"

"Oh, is that the gangster one?" said Maarten. "I love that gangster one!"

"No, you're thinking of 'A Piece of the Action.' Where Kirk and Spock land on this planet full of Chicago gangsters."

"Right!" Maarten said. "Spock in a gangster suit!"

"Yes, they called him Spacco," Don said. "But actually, the one I was thinking of is 'The City on the Edge of Forever.' Where they have to investigate this time disturbance. And Kirk runs through this doughnut."

"Ah—do I remember that one?" Maarten wondered.

"And Kirk ends up in New York during the Depression," Don said. "With this fantastic woman—"

"Yes!" Maarten said.

"Joan Collins!"

"Right!"

"Kirk falls in love with her, and who wouldn't. But it turns out she's the leader of a pacifist movement and that she is going to prevent the United States from entering World War II. Then Kirk finds out that she is going to be hit by a car. He could save her, but he has to let her die."

Jim Gunn leaned back in his chair and closed his eyes.

Later Don Schneider said, "I think we need music for this movie."

The night assistant agreed. Juan Carrasco crossed the room and turned on the stereo. Instead of music, he got the news: "The price of gold jumped thirty-five dollars an ounce today, in hectic trading."

The announcer moved on to the local headlines. Apparently reading from hot copy, he continued: "A San Diego man was indicted today on charges of—oh. Uh—sexual things."

The astronomers were startled. "Juan! What is that?"

"I couldn't say."

Juan turned the dial to KFAC Los Angeles, and Vivaldi's *Four Seasons* came up, conducted by Seiji Ozawa. "Spring" added a touch of grace to the galaxies streaming past, as the violins called to each other like birds. The screen filled with blotty things, a snow flurry. Suddenly at least two hundred galaxies spattered the screen. The core of an unnamed, unknown supercluster of galaxies was coasting through the Hale Telescope's field of view.

George Ellery Hale was a solar astronomer who invented clever machines for studying the sun. He also had a talent for extracting money from tycoons, which he applied to the creation of telescopes. He had a powerful imagination. It overpowered not only him but also everyone around him; and as he talked and wrote of bigger telescopes, he drew businessmen, politicians, and scientists into his precognitions of giant glass. First came the forty-inch Yerkes Telescope, at Williams Bay, Wisconsin, finished in 1897—a refracting telescope that uses lenses to gather light. A Chicago industrialist and expert stock-waterer named Charles Yerkes paid for it. In its day it was the largest telescope on earth. The next of Hale's telescopes was the sixty-inch reflecting telescope, finished in 1908, which is on the summit of Mount Wilson, overlooking the city of Pasadena. Hale's father paid for the mirror, while Andrew Carnegie paid for the rest of the telescope. For a time it was the largest on earth. The third was the one-hundred-inch Hooker Telescope on Mount Wilson, also once the largest on earth and mostly paid for by a Los Angeles hardware king named John Hooker. Hale became the director of the Mount Wilson Observatory, located on Santa Barbara Street in Pasadena.

Hale paid for his telescopes with broken nerves. He suffered from a New England family background, a culture in which nervous and physical diseases can no more be separated from one another than a photon can be separated into a wave or a particle. Hale radiated a boyish energy through oval spectacles, and ladies found him charming. He used to jog up Mount Wilson along miles of steep switchbacks, reciting Italian poetry. When in the grip of his overheated imagination, Hale had a peculiar habit of holding his hands rigidly outward from the middle of his body and gazing into the distance. He had a slight body that seemed always in motion, except when it was confined to bed, which was often. He suffered from violent headaches, physical prostration, frantic excitement, insomnia, ringing ears, tingling feet, indigestion, and a general sense that his mind was whirling out of control. Later in life he gave this galaxy of symptoms different names. He called it the "Americanitis," because he felt that Americans had a tendency to let the wings of their ambition drive them to insanity. Or he called it "the whirligus." One night, when Hale was forty-two years old, he was sitting up in his bedroom enduring the whirligus when a little man materialized before him. That was the first appearance of the elf. The elf gave Hale some advice on how to run his life. Hale appreciated the advice and thanked the elf, and the elf went away. When (in succeeding months) the elf started coming back, Hale began to worry. Hale would hear a ringing in his ears, heralding the arrival of the elf, who would materialize and deliver some advice. Hale was naturally reluctant to tell his family and friends about the elf, even when the elf began to follow Hale around during the day. At night Hale would walk in circles around his bedroom under the spell of malignant dreams, during which, according to one of his friends, "in his tormented half-sleep [he] would try to climb the picture frames on the wall." The elf may have urged Hale to seek psychiatric help. In any event, every now and then Hale would buy a train ticket to the East Coast and check into a sanatorium in Maine, where he would stay for a few months. There he would saw and split a few dozen tons of wood in order to calm himself. He bought a three-wheeled motorcycle. Riding on it through Pasadena one day, he noticed two men on motorcycles. He shouted at them, "Want to race?" and gunned his motorcycle. Then he heard the sirens—he had failed to notice that they were policemen. He tried to give the cops the whirligus, but they arrested him, the director of the Mount Wilson Observatory. In 1922, Hale's nerves became so bad that his doctors feared he was heading for a major nervous breakdown. They persuaded him to take a trip abroad for his health. He chose to go to Egypt.

With his wife and children, he sailed up the Nile in a lateen-rigged yacht, to the Valley of the Tombs near Luxor, where Howard Carter was just then opening the tomb of Tutankhamen. Hale visited the excavation and watched while the archaeologists pulled a procession of golden treasures from the underground rooms. Tutankhamen had been the son of the Pharaoh Ikhnaton, who had attempted to establish the worship of the sun god, Aten, throughout Egypt. Hale's biographer, Helen Wright, in her book *Explorer of the Universe*, said she believed that the sight of King Tut's tomb may have produced effects in Hale opposite from what Hale's doctors had hoped when they had urged him to travel. Death, eternity, and the Nilotic sun festered in Hale's brain, until all he could do was sit in a shady corner of his yacht and stare across the river at yellow cliffs, which to Hale seemed "pierced with the doorways of rifled tombs," as he wrote in a letter. Hale's resignation as director of the Mount Wilson Observatory soon followed.

When he returned to the United States, he went into partial seclusion at his home in Pasadena, to carry on his own researches into the sun. He built a private solar laboratory on a piece of land that had once been part of the original grounds of the Henry Huntington estate, and equipped the building with mirrors. The building's doorway was of carved wood and stone, arched over with hieroglyphs and an image of the rayed sun, copied from a tomb at Thebes; and the building had an underground room where Hale brought a shaft of sunlight into his instruments, and where he hoped to spend the rest of his days. But he could not keep his mind from a mirror, around which his imagination had been turning since the end of the Great War. Hale had not severed his contacts with friends and colleagues, nor with the Mount Wilson Observatory. He took the oxymoronic title of Director Emeritus, which suggests that some kind of force-field emanated threads of energy from his solar laboratory across the city of Pasadena to the headquarters of the Mount Wilson Observatory on Santa Barbara Street. In fact Hale continued to be a strong influence there during the early 1930s. Hale, in turn, may have been under the influence of the elf. Helen Wright, Hale's biographer, learned about the elf from a certain Dr. Leland Hunnicutt, who was a friend of Hale and who had listened sympathetically to Hale's descriptions of the elf. Wright does not say exactly what kind of advice this elf was giving to Hale, although she says that the elf "became almost a mascot." For all we know, the elf may have had in mind a two-hundred-inch telescope, which raises the possibility that one of the great scientific instruments of the twentieth century may have been built partly on the advice of an elf.

When Hale felt well enough to face the rigors of contact with humans, he received visitors in the library of his solar laboratory. There, stabilized in an armchair that had a book rest and a writing table attached to it, beside a fireplace that displayed a bas-relief of Aten riding his chariot into the sun, Hale talked and wrote about a giant mirror until he gave everyone the whirligus. At that time the one-hundred-inch Hooker Telescope on Mount Wilson was the world's largest. Hale pointed out that a mirror two hundred inches in diameter would have *four* times the surface area of a hundred-inch mirror; four times the light-gathering power. "Starlight is falling on every square mile of the earth's surface, and the best we can do at present is to gather up and concentrate the rays that strike an area one hundred inches in diameter," he wrote in 1928, in an article for *Harper's* magazine. These words struck a chord with the board of trustees of the Rockefeller Foundation, which after consulting with John D. Rockefeller, Jr., decided to provide funds for Hale to build a two-hundred-inch telescope.

Hale naturally wanted the money to go to the Mount Wilson Observatory. But the Mount Wilson Observatory was funded by the Carnegie Institution of Washington, which had been endowed by Andrew Carnegie. The Rockefeller trustees did not like the idea of giving Rockefeller money to a Carnegie observatory. The resulting negotiations nearly pitched Hale into bed, but he managed to work out a compromise. The Rockefeller Foundation gave six million dollars to the California Institute of Technology, in Pasadena, while at the same time Hale secured an agreement between the California Institute and the Mount Wilson Observatory to cooperate with each other in the building and operation of the telescope.

When the money had been pledged, Hale established various committees to plan the telescope. John Anderson, a Mount Wilson astronomer, was appointed executive officer of the project. Various sites on mountains throughout southern California were tested for dark skies and good seeing. "Seeing" refers to turbulence in the atmosphere. Viewed through a large telescope, stars appear to waver and tremble, as if they were near a radiator. Poor seeing causes stars to twinkle. Hale wanted to find a mountain where the stars did not twinkle. In the spring of 1934, Hale and Anderson drove up Palomar Mountain in a Pierce Arrow touring car, along a winding dirt road. (Two years earlier, on the occasion of Albert Einstein's first visit to Pasadena, the Mount Wilson Observatory had blown a small fortune on this automobile, in order to put Einstein in a luxury car.) Hale and Anderson went from one end of the long mountain to the other, comparing sites for the telescope, until they chose a fern meadow at an altitude of 5,600 feet. The meadow was accessible, yet far from city lights, and something about the meadow's topography muted the air above it, calming the stars into points during many nights of the year. The site had water too—a ravine opened to the north, where Horse Thief Spring trickled from beneath oaks that had been growing a century before Isaac Newton was born. The San Luiseño Indians had called the place *Poharup*—Noise of Falling Water.

The major responsibility for designing the telescope went to John Anderson and a committee of astronomers and engineers. They tested and discarded a number of designs for the tube and mounting of the telescope before they hit upon what is now known as a yoke-and-horseshoe mounting. The telescope's tube swings between the arms of a fork, which resemble the arms of a tuning fork and are called a yoke. As the tube tracks the stars from east to west, the tuning fork rotates around its handle; the arms of the yoke turn. The tube of the telescope weighs heavily upon the yoke, and so the ends of the yoke rest for support on a gigantic horseshoe bearing—a lazy C, which slides on its back, floating on a layer of Flying Horse telescope oil. The tube is a scaffold made of I-beams, fifty-five feet long, designed by a structural engineer named Mark Serrurier. Serrurier's design is called the Serrurier Truss. The truss flexes under strain, as does a bridge. Both ends of the tube are free to sag by up to a quarter of an inch, and yet the tube holds the two principal mirrors—the sixteen-foot primary mirror and the four-foot secondary mirror at the top of the telescope—in perfect parallel alignment to within a hundredth of an inch, no matter where the telescope points, and thus keeps the stars in focus. "I was given a job nobody thought could be done," Mark Serrurier told me. "That's where I got my satisfaction."

During the summer of 1936, work crews blasted and dug a circle of holes in the fern meadow. Caltech students hauled rock out of the holes by hand, using wheelbarrows. The dome was designed by committee. Russell Porter, an artist, explorer, and amateur telescope maker, may have elaborated some of the Art Deco decoration on the dome, although even these details seem to have been worked out by the committee. Porter noticed that the dome's size was within two feet of the diameter and height of the Pantheon in Rome. The committee had evidently not planned that.

The Westinghouse Electric and Manufacturing Company, in South Philadelphia, cast and machined the tube, the yoke, and the horseshoe bearing. A workman named William Ladley put the last rivet into the Serrurier Truss, before a crowd of dignitaries in South Philadelphia, including Albert Einstein. Tube, yoke, and horseshoe bearing traveled to California through the Panama Canal, chained to the deck of a freighter. Those parts were assembled on the mountain, inside the dome, under the direction of a master engineer named Byron Hill, who later became the observatory superintendent. I found Byron Hill in a double-sized mobile home on top of a hill in Tuolumne, California, with his wife, who was in poor health. He spends his mornings feeding birds on the patio and drinking coffee, and does not spend a lot of time congratulating himself on what he did to improve the vision of the species. "I get a little older every day," Byron Hill said. "I object to it." During his days as superintendent of the observatory, the astronomers sometimes referred to Palomar Mountain as Byron's Hill. They regarded him as a tough customer; he used to wear a leather jacket and aviator's glasses. He once threw an astronomer out of the dining room for wearing Bermuda shorts—"His legs *shocked* the housekeeper," he explained to me. On another occasion a night assistant parked his truck inside the Hale dome, where Byron thought it did not belong. Byron fitted a chain around the truck, hoisted the truck up to the top of the dome, and let it dangle next to the Hale Telescope. About the tube, yoke, and horseshoe bearing he said, "The things fitted together beautifully."

The mirror-blank was cast at Corning Glass Works, in Corning, New York. George McCauley, a master of Pyrex, directed the work. McCauley was a taciturn man. Asked how he planned to cast the glass, he said, "It will be no different than making a bean pot, except in the methods employed." The methods included building an igloo-shaped oven and casting a series of disks inside the igloo, in molds that resembled waffle irons. McCauley started with small disks and worked his way up to two hundred inches. His methods produced a mess during the first casting of a two-hundred-inch disk, when pieces of the mold broke off and floated away in a mulligatawny of hot Pyrex. Asked what he planned to do next, McCauley snapped, "We'll just make a new disk." On December 2, 1934, McCauley's men ladled about forty buckets of white-hot Pyrex into another waffle mold. The Pyrex was thick stuff and oozed out of the buckets in glops, like refrigerated honey. McCauley kept the melt in the oven for ten months, gradually cooling it, letting the glass anneal.

When McCauley opened the cold oven, he saw that he had made the world's largest monolithic piece of glass. It had a hole in the center, like a doughnut. Corning engineers encased the disk in a steel shell and stood it upright on a flatcar, to be drawn by a steam locomotive to California. For more than two weeks the telescope train crossed the United States, often at speeds of five miles an hour. Every time the train stopped, armed guards dived underneath it, looking for hobos trying to ride under the disk, for this was the Depression. Huge crowds turned out. Ten thousand people in Indianapolis watched the telescope train pass. Afraid that someone might try to take a potshot at the disk, Hale and Anderson felt it necessary to armor the disk with steel plates. If a bullet had broken the glass, that certainly would have killed Hale. At night the train was parked on a siding, illuminated with floodlamps, and patrolled by guards carrying loaded rifles, who had orders to let nobody approach within shooting distance. The train passed through St. Louis, Kansas City, Clovis, Needles, and San Bernardino, and arrived in Pasadena on Good Friday, April 10, 1936, witnessed by crowds. The disk was unloaded and lifted into the Caltech optical shop. A Pasadena newspaper reported: "There has not been such excitement since Ambler's Feed Mill burned."

The excitement was too much for George Ellery Hale. Too ill, physically and psychologically, to watch the triumphal entry of his glass into Pasadena, he had withdrawn from the world, broken on the wheel of whirligus. He spent his last years with his instruments in the underground chamber of his solar laboratory, looking into the sun. Day by day a heliostat mirror (a sun-tracker) turned slowly at the top of the building, throwing a shaft of sunlight into the basement, where Hale, staring through an eyepiece just two millimeters across, watched prominences heaving and lapsing around a ball of hydrogen as old as the world but never the same from one minute to the next. His grandchildren would visit him and listen to his stories, and perhaps the elf listened too. He maintained contact with the Palomar project through long letters to a few friends. In 1938, at the Las Encinas sanatorium in Pasadena, he said to his daughter, Margaret Hale, "It is a beautiful day. The sun is shining, and they are working on Palomar." He died a few days later. Hale had not returned to Palomar Mountain since the day he chose the fern meadow. He never saw his greatest telescope.

Marcus Brown, Caltech's chief optician, directed the grinding of the mirror. Brown hired twenty-one unemployed men (mostly right off the street) to operate a polishing machine. Brown's men wore white suits and white sneakers—clothing that never left the shop. The glass disk sat on a turntable. While the turntable rotated, an arm pressed a rotating circular polishing tool against the glass; the arm moved the tool in differing directions across the glass, thus tracing overlapping cycles of movement known as Lissajous figures.

I drove up into the Verdugo Hills, near Pasadena, one afternoon in spring, along an unmarked dirt road, until I found a sunny house where lived Melvin Johnson, who as far as I could tell was the only master optician still alive who had worked on the two-hundred-inch mirror. We sat and drank coffee, and Johnson said that it had been so long since he had talked about that mirror that he might have a little trouble finding the right words, but then his words began to move in Lissajous figures around a giant disk of flame Pyrex with a hole in the middle. The opticians inserted a Pyrex plug into the hole before they started to polish the disk. The polishing tool was covered with a layer of black pitch, which rubbed and sleeked against the glass. The formula for the pitch changed now and then, Mel Johnson said, and the method of cooking the pitch in a pot was essentially a black art. "We tested all kinds of mixtures. I threw out a garbage can full of formulas," he said. The pitch, he said, contained amber rosin from Alabama pines, pine-tar oil, and beeswax. Hoping to get a smoother polishing action on the glass, the opticians experimented with pitches adulterated with paraffin wax, automobile motor oil, and a powder made from ground walnut shells, "which was like flour," Johnson said. Every few minutes the opticians poured across the glass a slurry of water and Carborundum grit. They used finer and finer grades of Carborundum and then switched to red jeweler's rouge. By 1941, they had polished away five and a quarter tons of glass, had used up thirty-four tons of abrasives and jeweler's rouge, and had brought the surface of the Pyrex disk down to a hollow sphere. From there they had to deepen the glass slightly into a paraboloid. A paraboloid is a saucer that focuses light to a point. The layer of glass that they had to remove in order to parabolize the mirror equaled the thickness of half a human hair. This work required eight more years of polishing, interrupted by World War II, when Caltech halted work on the telescope.

The opticians were afraid that their machines would drop a metal filing on the glass. A grain of metal or grit trapped between the polishing tool and the glass would have cut a helical scratch in the glass that would have delayed the project for six months, or perhaps for years. They swept the room with vacuums and electro-magnets. Then they looked at the dust they had collected under a microscope, classified the particles, and saved them in envelopes. If they saw a dust particle of a type they did not recognize, they stopped all their machines until they could trace the particle to its source. Toward the end of the polishing, the opticians spent more time testing the glass than rubbing it, fearful that they might polish too deeply in places, especially around the outer edge of the glass, in which case they might never be able to resurrect a true optical surface. Their testing apparatus was keen enough so that an optician could place his hand on the glass for a minute until the glass warmed, take his hand off it, and see a swelling in the shape of a hand persist on the glass. Before they looked at the glass through the testing apparatus, they had to turn off all fans and prevent people from walking around the room, "because a current of air coming through the room made the air look like a smoke screen," Mel Johnson said. He remembered seeing waves twitching along the surface of the glass, as if the glass were restless, gently pulsing with life. The waves mystified the opticians, until they discovered that the mirror was picking up harmonic vibrations from traffic on California Boulevard, near the optical shop. After that the opticians scheduled precision testing of the glass for early Saturday mornings.

When the surface of the glass had reached a fairly acceptable paraboloid, the opticians removed the plug from the hole in the center of the glass. In November 1947, they mounted the glass in a steel mirror cell (it would never leave the mirror cell again) and put it in a box and carried it in a flatbed truck up Palomar Mountain. The purpose of the superstructure of the telescope is to move the glass around and to keep it pointed at one spot in the sky. The purpose of the glass is merely to support five grams of reflective aluminum in a perfect paraboloid, in order to focus starlight into a camera. John Strong, a physicist, had invented a technique for depositing aluminum on glass. Strong had taught the Caltech opticians his trick and then moved on. He eventually wrote a textbook on physics. When I asked around Caltech about John Strong, people seemed to think that he was dead. I made some telephone calls to various parts of the United States and turned up John Strong in Amherst, Massachusetts, nowhere near dead, because he was working on a new edition of his textbook. "I never saw the mirror again," Strong said over the telephone. He explained that he had had to clean the glass in order to make the aluminum atoms stick to it, for he had learned that oil from the human skin, which inevitably got on the glass from the opticians' hands, caused aluminum to crinkle off. Strong had tried washing astronomical glass with chemical solvents, but no solvent seemed powerful enough to remove skin oil. Then Strong discovered Wildroot Cream for the hair. "I never used it on my own hair," he said, "but it was one of those things you just knew about." He mixed powdered chalk with Wildroot Cream and rubbed it all over the two-hundred-inch glass, which terrified the opticians. "In order to get glass clean,"

Strong told them, "you first have to get it properly dirty." He wiped the sludge off with wads of felt, leaving a molecular film of Wildroot Cream on the glass. He placed the glass in a vacuum chamber, then fired hot electrodes over the glass, which burned off the Wildroot Cream along with the fingerprints, leaving virgin glass. "Wildroot Cream was one of those little black arts," Strong explained to me. "It has Peruvian lanolin in it." While the glass was still sitting inside the vacuum chamber, Strong vaporized aluminum wires in the chamber, and aluminum fell in a dew over the glass.

The opticians opened the tank; the glass had become a mirror. Three nights before Christmas, 1947, a crowd of astronomers and engineers gathered in the dome for first light. They rolled the mirror cell under the butt of the telescope. They raised a hydraulic jack and inserted the mirror into the telescope. Byron Hill's workmen began tightening a circle of bolts around the mirror.

A bang and a hideous squeal filled the dome. It sounded like a pig being clubbed to death—the unmistakable screech of a crack fingering through sixteen feet of Pyrex. Many eyes turned toward John Anderson, who had been waiting twenty years for this moment and who had a heart condition. After a silence during which Anderson did not collapse, a workman said, "You ever seen a one-million-dollar bolt snap?" The bolt had not snapped off, anyway; it had only creaked. A few minutes later, John Anderson sat in a lift chair, which raised him fifteen feet until he could peer into an eyepiece mounted at the base of the Big Eye. He gazed for a while into the Milky Way, in silence. When he came down, somebody asked him, "What did you see?"

"Oh, some stars," he said.

One by one the astronomers and engineers sat in the lift chair and went up to the eyepiece. When Byron Hill got his chance to look, as he would remember, "I had never seen so many stars in my life. It was like pollen on a fish pond." The sight, he said, "made me feel pretty good."

They all knew that much tweaking and polishing still remained to be done. First light on a large telescope is the beginning of a process of adjustment that may continue for years. Although glass is brittle, it is actually a supercooled liquid. Glass is physically similar to Jell-O. Glass can flop, tremble, and shudder. As a large mirror moves through varying angles, it buckles and droops. The Hale Telescope's mirror is rubbery. You could push down firmly on it with your thumb and throw the stars out of focus.

Today, pressure pads controlled by computers push and warp large telescope mirrors to keep them in shape. When Hale first proposed a two-hundred-inch mirror, he sensed that the problem of supporting a lake of glass to a tolerance of four millionths of an inch across two hundred and nine square feet of surface area might be impossible to achieve with existing technology. He decided to hope that the technology would come along. In the early 1930s, an engineering team designed and built thirty-six mirror-support machines, weighted with lead. When the glass disk arrived in the Caltech optical shop, the machines were plugged into pockets in the back of the disk. An engineer named Bruce Rule then tested the glass for signs of slumpage and found that the glass behaved somewhat in the manner of uncured latex rubber—when the opticians leaned the mirror at an angle, the glass would droop and not return to normal shape for quite a while. The mirror-support machines were failing to compensate for slumpage in the glass. During the summer of 1948—six months after first light—Bruce Rule extracted the mirror-support machines from their pockets in the glass and rebuilt the machines. Rule's thirty-six mirror-support machines work passively, by means of levers and lead weights. The levers barely move, yet they exert three-dimensional forces throughout the glass, which, in places, reach stresses of up to twelve hundred pounds.

Bruce Rule was a tall, white-haired man who wore thick glasses and spoke in a soft, measured voice, and who was widely regarded around Caltech as a genius, which is a reputation not easy to get in a place like Caltech, where the geniuses do not generally refer to each other as such. I visited Bruce Rule one day at his home in Pasadena. "I wouldn't call them machines," Rule said. "I would call them compound support units." Each unit, which resembles a piston inserted in the glass, contains an uncounted number of parts. Rule said, "I think that between six hundred and one thousand parts in each unit is a reasonable number." Since there are thirty-six mirror-support units, that would mean that the Hale mirror is held up by as many as thirty-six thousand pieces of metal, most of which move, if only slightly. Now we see why Bruce Rule was considered a genius. Rule said, "That estimate depends on how you want to count parts. If you want to count all the little parts inside ball bearings, then the number would be larger." The support units are, in fact, mechanical computers. They react to forces in the mirror and apply corrective action. Rule said, "I never recommended that this type of system be tried again." Virtually everybody at Caltech understands electronic computers, but nobody at Caltech understands mechanical computers, and consequently nobody dares to monkey with Bruce Rule's support units. Since 1948, there has been one attempt to oil them. It was not much of a success. The lead weights on the units are adjustable, but nobody wants to adjust them. Once or twice a year an engineer walks around the cage at the base of the telescope and reaches up inside the mirror cell. He takes hold of the weights and wiggles each in turn, in order to give the units a bit of exercise; but the feeling around Caltech is that anybody who tries to open Rule's units to see what is inside them will get himself fired. Rule did not worry about his units. "We didn't give ninety-day guarantees," he said. "We built for life."

Once in a while these days, the stars on the video screen turn into hollow triangles—the support units have become stuck. The astronomer turns to Juan Carrasco and says, "The mirror needs exercise." Juan then slews the telescope from horizon to horizon, from north to south, from east to west, until the stars turn back into points. The nightmare of the engineers who take care of the Hale Telescope is that one night the stars will turn into triangles, Juan will exercise the mirror, and the triangles will get bigger. In that event, the engineers would have to search the Caltech archives for microfilm of Rule's blueprints for the support units, although no Caltecker is sure that he would understand the blueprints. During the summer of 1948, when he was designing the units, Bruce Rule liked to go to the beach for a weekend, where he would lie on the sand and hear the surf and see shapes in his mind's eye—levers and pistons and rippling glass. "I could keep a crew of thirty draftsmen going," Rule said.

About the time that the mirror went into the telescope, the chief optician, Marcus Brown, retired. An astronomer named Ira Bowen was named director of the observatory, and Bowen took personal control over what is known as the final figuring of the mirror. In the spring of 1949, the opticians stripped the aluminum from the glass and went to work on the glass with small polishing tools. Under Ira Bowen's gaze an optician named Don Hendrix did much of the polishing; Melvin Johnson assisted Hendrix. They would mount the glass in the telescope. Bowen would look at a bright star, which he could see reflected in the bare glass, and take measurements of the star, while Hendrix or Johnson would outline any defective zones in the glass, using a grease pencil. These measurements would require from one to three nights to complete. Finally, at dawn, they would remove the glass from the telescope and rest it on a cradle. Hendrix and Johnson would rub the glass in one or two spots, using their polishing tools. Hendrix's favorite tool was an aluminum disk faced with black pitch, the size of a Thin Mint. Sometimes they used a piece of cork. The smallest defective zones were an inch or two across and circled with a grease pencil. To polish these zones Mel Johnson would dip a watercolor brush in water mixed with a silt of polishing compound called Barnesite. He would paint the zone with Barnesite, then rub the zone with his thumb. "There are no sharp edges on your thumb," he said. "Your thumb flows into the zone." Johnson liked to use his thumb because he could feel the temperature of the glass change as he rubbed it. Each stroke removed about two hundredths of a millionth of an inch of glass, but heat from the rubbing swelled the glass by more than that. They would polish a little here, a little there, until they sensed that they had swelled the glass. Then they had to let the entire two-hundred-inch glass cool for the rest of the day, to let the swelling go down, before they could see what they had done to it. They would mount the glass in the telescope and test it on a star. This process continued throughout the summer and fall of 1949. "All we were asking for," Johnson said, "was a hard point of light." In the end they polished the mirror to a mathematical formality. If the Hale mirror were expanded to the size of the United States, it would exhibit no hill higher than four inches. That is not counting pits left by bubbles in the glass, which the opticians plugged with pitch.

Bowen's final tests revealed an astigmatism in the glass—the mirror was slightly warped. The opticians could have polished the glass for another three years, but they solved the problem with a kludge. They purchased four fisherman's scales at a dime store and hooked the scales to the back of the glass, where their springs each tugged at the mirror with about seven ounces of pull, just enough to open the mirror by a few millionths of an inch and flatten the warp. In 1981, a Caltech engineer, poking around nooks in the back of the mirror cell, asked himself what in the world these fisherman's scales were doing in there, and removed them. When the astronomers complained that the telescope had gone out of focus, the fisherman's scales went back on the mirror in a hurry.

There is a saying among those who polish astronomical mirrors for a living that an optician never finishes a mirror—you take it away from him. By the autumn of 1949, Hendrix and Johnson were still rubbing the glass with pitch, cork, and thumbs. They kept telling the astronomers, "We'd like another week." That would provoke a stormy meeting behind closed doors, because the astronomers were hot to use the telescope. "Ira Bowen just kicked all the astronomers in the teeth," Byron Hill said. Mel Johnson said that, for his part, he could have lived happily with that glass for another two years. The astronomers began to go wild. Bowen kicked them in the teeth again, and they came back like dogs for more. Bowen finally gave up. He took the mirror away from Hendrix and Johnson—he declared the mirror finished. He ordered Hendrix to put a coating of aluminum on the mirror and to put it in the telescope. In November 1949, the Hale Telescope went into regular use.

George Ellery Hale never knew that his telescope would be called the Hale Telescope, for the naming of the instrument did not occur until a dedication ceremony in the dome on June 3, 1948, more than ten years after Hale had died. James R. Page, chairman of Caltech's board of trustees, opened the dedication with a speech in which he said, "This telescope is the lengthened shadow of man at his best." One wonders what the elf thought of that remark. Later Bruce Rule slewed the telescope back and forth over the heads of a mass of spectators, while he prayed to heaven that the telescope would not drop any nuts or oil into the crowd.

Every six months Palomar engineers remove the mirror from the telescope and wash it with natural sponges, using Procter & Gamble Orvus soap, which is Ivory Soap without fragrance. Once a year the engineers strip the aluminum off the glass. The superintendent of the observatory is a stocky man with a mustache, named Robert Thicksten. After the aluminum is stripped off, Thicksten stands on a little platform in the hole at the center of the glass and inspects the bare glass. The glass, Thicksten said, reminds him of a jewel. George McCauley's flame Pyrex holds lapidary colors that change under different lighting. Sometimes the glass seems to be topaz yellow, or pale green, or amber, like maple syrup. Under bright lights, the glass discloses a ring of blue brilliant haze, caused by an unknown contaminant in the Pyrex, encircling the hole in the center, as if the glass had a blue iris. From above, the waffle structure in the back of the glass is clearly visible—a network of triangles and hexagons that gives the glass the frightening appearance of an insectiform eye. Backlighting reveals dark shapes caught in the glass—chunks of firebrick that had broken away from the oven and dropped into the melt. The glass sparkles with silvery air bubbles and is swirled with liquescent folds, eddies, and stria, which, in one area, erupt at the surface of the glass into a spiderweb of cracks. The cracks hold pockets of red jeweler's rouge that worked its way downward into the glass. The opticians had drilled small holes at the ends of the cracks, to arrest the cleavage through the glass, and plugged the holes with pitch. The cleverest human craft had barely harnessed the natural world's random currents in the two-hundred-inch mirror, a surface for gathering light, intermediate between land and sky, that had risen in a dream before the eyes of a frail dreamer, and which the work of almost anonymous human hands had made into a physical thing. The inspection complete, Bob Thicksten and his engineers clean the glass with solvents and put it in a vacuum tank and vaporize aluminum over it, which turns it back into a mirror. Thicksten has elected to stop massaging the glass with Wildroot Cream. "That was a black art," Thicksten believes.

Virtually all of the Hale's builders are gone. John Anderson died of a heart attack in 1959. George McCauley, Russell Porter, Marcus Brown, Ira Bowen, and Don Hendrix are dead. Over the years the collaboration that George Ellery Hale had negotiated between the Mount Wilson Observatory and Caltech became more delicate, more formal, more punctilious, and exploded in a cat-fight—the sort of thing that happens all the time around telescopes. In 1979, Caltech assumed administrative and financial control over the telescopes on Palomar Mountain, including the Hale. The Hale glided through the divorce. It had been so beautifully engineered that it seemed to run according to a stubborn will of its own, aloof from human frailty, but if it ever broke down in a major, unforeseen way, Bob Thicksten believed that about six people on earth would know how to repair it. Or *might* know. On summer nights Thicksten would stand on the catwalk of the dome, listening to the nocturnal sounds of the Big Eye, asking himself if its gears were humming on the right note. "We know when certain things work," Thicksten once remarked to me. "But the worry is, we don't know *how* they work." The Big Eye had outlived its creators.

▪ ▪ ▪

The Hale's builders equipped it with a number of smaller mirrors so that light can be directed from the main mirror to various observing stations. One station is a small room at the top of the telescope called the prime focus cage. An observer can sit in prime focus and look through an eyepiece directly down into the main mirror, where the observer sees a reflection of the deep. When I consider examples of the power of the Hale Telescope, what comes to mind is a story that Don Schneider told me about something that happened to him one night when he was working in prime focus. Near dawn he had a few minutes of extra time on his hands. He had never seen Venus through the Big Eye. "Point me to Venus," he said to Juan Carrasco over the intercom. The prime focus cage tilted as Juan slewed the telescope down and east, toward the horizon, where Venus was. "We are there," Juan said. "You have two hundred inches on Venus." Don started to look into the eyepiece. A stab of pain hit him in the eye. He pulled his head back, and a pencil-thin shaft of white light came out of the eyepiece. The light was too bright to look into, and it reminded him of a movie projector beam. It was the light of Venus that had fallen on 209 square feet of mirror and had been condensed into the eyepiece. He could see dust motes dancing in the light of the morning star.

George Ellery Hale's greatest telescope is a time machine. It reimages lost time. Light from the sun takes eight minutes to reach the earth. The sun is eight minutes into lookback time. Photons from Venus take anywhere from two minutes to fourteen minutes to reach the earth, depending on where Venus is in its orbit in relation to the earth. The planet Saturn is one light-hour away. Proxima Centauri, the nearest star other than the sun, is currently about four light-years and three light-months away from the earth. (Proxima Centauri is moving, and someday it will be nowhere near us at all, because stars are lone voyagers through the galaxy.) A few dozen known stars now drift near the sun, bearing names such as Epsilon Indi, Tau Ceti, Kruger 60, Kapteyn's star, and Procyon. Distant, giant stars—Rigel, Aldebaran, Betelgeuse, Antares—are hundreds of light-years away. The mist in the Milky Way consists of stellar images that are thousands of years into lookback time, because those stars are, on average, a few thousand light-years away from the earth. The Milky Way appears to be a band of light encircling the sky, because the Milky Way is a spiral galaxy shaped like a disk, and we are within the disk, looking outward. A spiral galaxy is a rotating cloud of matter that contains much gas and dust, a prodigious quantity of a stuff known as the dark matter, which astronomers admit they know almost nothing about, and about one hundred billion stars. A number like one hundred billion is not easy to imagine. If you placed that many ten-dollar bills end to end, they would form a line of bills that would go eight times around the earth, and then out to the moon, back to the earth, and out to the moon again. (We need to elect some astronomers to Congress.) Two square miles of growing wheat contain roughly one hundred billion grains of wheat. A star is to a galaxy what a grain of wheat is to a farm in Kansas.

Consider a sun the size of the dot over this *i*. On that scale the earth would be the size of a one-celled microorganism, located about two inches away from the sun. On that same scale the nearby star Proxima Centauri would be about nine miles away—and the center of the Milky Way would be about fifty thousand miles distant. If something were to happen to the earth, it would not be missed. Man is dispensable. So is the earth. The one hundred billion stars in the Milky Way, including the sun, revolve around the Milky Way's center, just as the earth revolves around the sun. The sun and the earth take about 250 million years to make one orbit around the galactic center—a period of time known as one galactic year. The sun and the earth have existed for about eighteen galactic years—they have traveled about eighteen times around the galaxy since they were formed. Some kind of extremely heavy, compact object is sitting at the rotational center of our galaxy and giving off radio waves. The radio signals from the galactic core that we are now picking up began to travel toward us in about 23,000 B.C., around the time that Upper Paleolithic hunters were painting handprints on the walls of caves in the Pyrenees.

Not too far from the Milky Way float other galaxies—the Clouds of Magellan, the Draco dwarf, the Fornax dwarf, Andromeda, the Pinwheel, the Whirlpool, Centaurus A, the Sombrero, the Zwicky Antennae, Stephan's Quintet. Andromeda, a near neighbor, is a spiral galaxy about two million light-years away. If the Milky Way were the size of a dime, then the Andromeda galaxy would be another dime about two feet away. Certain mysterious forces, which are not well understood, cause galaxies to evolve into extraordinary shapes: barred spirals, globes, footballs, rings, fuzz balls trailing rattails, thin smooth disks, and chaotic patches. Galaxies prefer company. They like to cluster in knots. A small knot such as the Local Group contains about a dozen galaxies, most of them dwarf galaxies, such as the Clouds of Magellan. A so-called poor cluster contains about a hundred galaxies. A rich cluster contains a few thousand interswarming galaxies.

Superclusters are the largest clearly identified structures in the universe. A supercluster is a megalopolis of galaxies containing dozens of rich clusters and uncounted thousands of galaxies, gathered into eddies or drifting alone. Galaxies that are not connected to any cluster are known as field galaxies, a term that somehow suggests the way in which wildflowers speckle a meadow. A typical supercluster has a peculiar shape: that of a ropy blob, like a yam; or maybe a curved sheet, like part of a bubble (astronomers are having a glorious time haggling over that). In any case, superclusters are cloudlike agglomerations of galaxies enclasping huge voids, or bubbles of apparently empty space.

The most densely populated region of our own Local Supercluster is a thick concentration of galaxies visible in the constellation Virgo. Those galaxies are maybe thirty, or maybe sixty, million light-years away from the Milky Way. Other nearby superclusters paint the sky: the Hydra-Centaurus Supercluster, the Perseus Supercluster. The Pavo-Indus Region is a supercluster containing hundreds of thousands of galaxies dusted across the southern sky. Then, as a telescope looks outward, the realm of the superclusters stretches into unmapped deserts of time. There may be about one million superclusters in the observable universe. These numbers are slippery. The universe may contain ten billion observable galaxies, or maybe one hundred billion—nobody really knows.

As a telescope looks backward into time (or out into space) the galaxies appear smaller and fainter. The atmosphere of the earth emits a slight natural glow at night, called the skyglow, which drowns the faintest galaxies. When a telescope probes about five billion light-years into lookback time, it can detect only the brightest galaxies—giant, elliptical galaxies—because spiral galaxies similar to the Milky Way are too dim to be seen at that distance, even with the best instruments. At extreme distances a telescope can only resolve the brightest beacons, the quasars. Quasars are the only class of luminous object that a telescope can see at cosmic distances, shining among and well beyond the most distant visible galaxies; and the most deeply redshifted quasars are probably the most remote objects that the Hale Telescope will ever detect.

During three nights in March, the scanning for quasars began. What these scans would reveal of the outer edge of the known universe remained unknown to Maarten Schmidt, who did not like to guess what surprises the universe might have in store for astronomers. The video screens were filled with galaxies, which were moving, giving the data room the appearance of the bridge of a starship. We must have been traveling at warp fifteen, except when 4-shooter bombed, when the screens would fill with jazzy stripes and we would drop to auxiliary impulse power, and Jim Gunn and Don Schneider would shout invective at 4-shooter and pound keys, while Maarten Schmidt whistled bits of Bach. When 4-shooter behaved, the astronomers liked to discuss what was passing on the screens.

Don Schneider touched the main screen. "Look at that, Maarten. A straight line of galaxies."

"It looks like a string," Maarten commented rather dryly.

"Goodness," Don said. "Still more galaxies. This must be a supercluster. There's lots of little junk going by." What he considered to be "little junk" were galaxies the size of the Milky Way, but viewed from such a distance that they were only dapples on the screen, like shoals of leaves that had fallen onto a pond.

I wondered aloud if the galaxies we were looking at had ever been given names by people.

Jim Gunn said, "Absolutely not."

"Have they ever been numbered or catalogued in some way?"

"No, actually not," he said.

"Have they ever been *seen* by any astronomers before?"

"I don't think so." Jim pulled a handkerchief and blew his nose. "Maarten, would these galaxies ever have turned up on a photographic plate?"

Holding a Chips Ahoy! cookie, Maarten reflected on that idea. "I would say not—eh what, James?"

"We are going pretty deep."

"Ja, except for the bright ones, most of these galaxies are too faint to show up on a survey plate."

"It is somewhat mind-boggling, isn't it?" Jim remarked. He turned to the night assistant. "This is an exciting night."

"Oh, yes," Juan said. "Everything is working."

Jim laughed. "Don't say that!"

The stereo in the data room was now playing Beethoven. While Maarten Schmidt found Beethoven not completely objectionable, he was mindful that J. S. Bach's three hundredth birthday was coming up. Crossing to the stereo, he said, "I'll interrupt this to see if there's Bach on the radio." He got a crash of cymbals and a soprano's wail. "That's not Bach." He kept turning. Human voices soared. Schmidt cranked up the volume. He had found a Bach cantata—and a radio station that was playing nothing but Bach that night. Schmidt said, "It's a good bet that you can find him on the radio on the night before his birthday." A while later, the B Minor Mass came on: "Gloria, Gloria in excelsis Deo . . . "

Juan leaned over and called to Don, "What do you think of the seeing tonight?"

Don thought it was pretty good on the whole, despite some high haze.

So did the Principal Investigator. He conducted the B Minor Mass with a Chips Ahoy! cookie held between thumb and forefinger, and the voices sang: "Et in terra pax hominibus /Bonae voluntatis . . . "

"Quick—Jim! There's a strange one!" Juan Carrasco called.

Jim Gunn rolled forward in his chair and stared at a large, bright galaxy. He said, "Is there a warp in that galaxy?"

Maarten Schmidt sat down and yanked off his glasses and squinted at the galaxy in question. It was bent like a crushed and twisted hat. Maarten groped around the table until he had found a plastic ruler. He put it against the drifting galaxy on the screen.

He said, "Why, yes, yes—ah, James—this does look like a warp."

"It's certainly not symmetrical," Jim remarked.

"Gad, that's a neat galaxy!" said Maarten.

"Isn't that beautiful," Jim said. "Someone could spend a long time studying that thing. Strange things . . . " His voice trailed off, and he took a swig from a can of Von's Lemon-Lime soda.

"Should I get a picture of it?" Juan asked.

"Yeah, go for it!" Jim said.

Juan grabbed a Polaroid camera from a shelf. He pointed it at his television screen and snapped a picture of the warped galaxy. He pulled the photograph and watched it develop. Slowly a contorted and apparently mangled galaxy appeared. This smear of light had met with an accident. Perhaps a heavy cloud of dark matter had fallen into it, or perhaps it had passed too close to another galaxy. Whatever the cause, a few tens of billions of stars had been sucked out of their normal orbits around the galactic center, thereby curling the galaxy like a bent bicycle wheel. Juan smiled. In all likelihood this galaxy had never been seen by human eyes before and might not be seen again for a long time. He said, "Beautiful, Professor James E. Gunn. You made my night." He aimed the camera at another galaxy. *Click. Zweee.* "Portrait of an Anonymous Galaxy." *Zweee.* "Still Life with a Galaxy"; "The Return of the Prodigal Galaxy"; "The Persistence of Galaxies"; "The Starry Night." A pile of Polaroids accumulated at his control desk. "You can have these," he said, pushing the pile at me. "I have too many already."

The Principal Investigator was less easily satisfied. Maarten Schmidt often went up to the catwalk that encircled the dome. He claimed that he was worried about the weather, but I noticed that the better the weather became, the more often Maarten Schmidt vanished to the catwalk and the longer he stayed there. I asked Jim Gunn what was going on. "Maarten likes to get dark-adapted," Gunn said, which I understood to be a polite way of saying that Schmidt had a peculiar habit of staring at the stars. When I asked Schmidt about this, his phrased answer was, "I find these trips to the catwalk not a tranquilizer, as it were, but a marvelous contrast to the pressures of the day." Given the slightest excuse, he would put on his parka and slip out of the data room. Walking underneath the Hale Telescope, he would cover his flashlight with his fingers, because the dome had to be kept pitch-black or the sensors in 4-shooter would go haywire. He would find a set of stairs, climb them, throw a lever, and open a steel door that led to the catwalk and the night sky. He would switch off his flashlight and stand in the darkness. Then he would walk slowly around the catwalk, traveling "anticlockwise" around the dome, as he described his preferred direction.

At fifty-five, Maarten Schmidt had reached the age at which prominent scientists can find themselves running their paperwork through a trash compactor in order to fit it in a briefcase. He served on a half dozen advisory boards and flew to conferences all over the world. He liked the catwalk at three in the morning, because, he said, "It is quite pleasant to be able to think about nothing in particular." He struck many American astronomers as an attractive yet distant and somehow unknowable figure, perhaps like a quasar. He was a familiar figure at conferences, where he seemed to make a passage through a sea of colleagues. He towered above them, distinguished by curly gray hair, a white shirt, and a bow tie—the president of the American Astronomical Society. He was born and educated in the Netherlands. After living for twenty-six years in southern California, he still carried a so-called green card, identifying him as an alien. He retained his Dutch citizenship and voted in Dutch elections. Unlike most astronomers, he dressed up for observing runs. He wore a checked sport coat and a shirt as red as a fire engine, and when he went up to the catwalk, he put on a dapper yellow cashmere scarf. Schmidt had international contacts. "I talk a lot on the telephone—too much," he said. "These days I find that I am doing all of my office work in the office and all of my science at home. Which is strange." After he left Caltech in the evening he had a quiet dinner with his wife, Corrie, sitting in their backyard. They saw the dusk settling and watched for the first stars. Strangely enough, the majority of professional astronomers do not know their constellations very well—they find stars by the numbers. But Maarten Schmidt knew his way around the sky. After dinner he would work on his quasars far into the night, and then he would watch television; later he would sometimes dream of the Big Eye, although he could never remember what happened in these dreams.

When he and Corrie had first been married, they had lived a casual existence, staying up until three in the morning and getting up late. Then they had had three daughters. Their daughters, Maarten explained, "didn't allow us to continue that way." Now that their daughters had grown up, Maarten and Corrie liked to take off for a little resort in the Anza-Borrego Desert, where they could sit outdoors in lounge chairs under a palm tree and look at the stars. They would pass a pair of binoculars back and forth, in order to discuss the excellence of a particular constellation. They had found a kind of peace watching things happen at night in the desert—jackrabbits running past, bats flipping after moths, meteors razoring the sky. They talked a bit or were quiet. Maarten particularly enjoyed the desolate yips of coyotes, a sound he had never known growing up in Holland. They noted the comings and goings of planets, and the galaxy wheeling overhead, until dawn caught them by surprise.

Maarten Schmidt grew up during the Second World War in the city of Groningen, in the north of Holland, where his father was a civil servant in the city government. He did not do well at sports (except at the high jump); he was the sort of kid who would rather look at stars. Groningen was blacked out during most of the war, and Maarten was thirteen years old when he first began to notice unnaturally brilliant stars hanging over his lightless city. They attracted him. His grandfather give him a thick magnifying lens and an eyepiece. Maarten taped the lens to a cardboard toilet-paper tube and put the eyepiece into the other end of the tube. He took the invention to the third floor of his house and looked out the window. He found a double star in Lyra. He explored the sky. Then the sirens would start up. Waves of Allied bombers passed over Groningen almost every night on their way to Hamburg and Bremen, and a tremendous roar of their engines shook the city. Sometimes the German antiaircraft batteries that ringed the city started shooting like mad, trying to hit the bombers, searchlights stabbing everywhere. Allied fighters would strafe the German batteries, or the bombers would loose their bombs into Groningen. One night a bomb fell in the street near their house. They huddled under the stairs until two in the morning, when the "all-clear" sirens sounded, and then they tried to get a little sleep.

Maarten's discovery in 1963, that quasars are brilliant and remote, had pushed him into fame, something he had not looked for and at times had resisted. Schmidt's quasars burned like beacons across unimaginable reaches of night. People would ask him if he had invented the Schmidt telescope. ("No—that was old Bernhard Schmidt. No relation. He was drunk most of the time, so he must have been brilliant in between.") His face appeared on the cover of *Time* magazine in 1967, when, after several years of photographing quasars alone in the prime focus cage of the Hale Telescope, he had smashed open the universe, had driven the limits of the Hale Telescope into territory beyond anything its builders had imagined, identifying quasars at greater and greater leaps, plunging into look-back time.

He lingered on the catwalk. So much about quasars remained unfathomable. In twenty-two years he had not found more than partial answers to his questions about their birth and death. In an interview with an historian of science named Spencer Weart, he once said that in his mind's eye he imagined science as a cloth being knotted together by many hands, in the manner of the anonymous Flemish weavers of old, who had worked side by side on benches. Corrie was a weaver. She had hung the inside of their house with large tubes and wheels of knotted cloth in muted colors. He lived surrounded by his wife's tapestries, which, he said, might have given him the feeling that science is a kind of tapestry extending back into the past. As he put it to the historian Weart, "I have strongly the feeling that as an astronomer on the earth, you are a link in history, because in science more than anything else, and certainly in astronomy, you build on what your predecessors did. You contribute a little here, you put in a couple of links there. It's all being knitted together, and a few of the stitches are yours." He sensed the presence of others working on the tapestry, sitting beside him; taking up threads, tying up small knots, while mysteriously, a design appeared. "And then," he said, "the fabric goes on." When your life as a scientist was over and others had taken up the weaving, you could always find your stitches in the cloth later, and you could say, "Well, they are there."

He wanted to know what had happened back in the early history of the universe, when quasars began to burn. He hoped that 4-shooter might see the light of the first quasars at the dawn of time; that 4-shooter might see first light. If these scans with 4-shooter could dig into time and bring up something new, then for Maarten Schmidt that would amount to a few more threads in a long cloth— a modest reply to nature's disregard for human reason. If the experiment worked, then 4-shooter would capture a few archaic photons of deeply redshifted quasar light. They would have left some remote quasar to travel through the void, without hitting anything, for almost as long as time itself had existed—a nice demonstration of how empty the universe is—until, some two or three times older than the earth, they ran into a mirror. The Milky Way lay in a mist along the eastern ridge of Palomar Mountain. Accustomed to the whine of the oil pumps, he noticed an eerie silence over Palomar; the oil bearings were shut off because the telescope was not moving tonight. Palomar Mountain was birdless and quiet. Even the toads had gone to sleep. One heard only a faint breeze fingering the cedars and humming along the catwalk rail. Coma Berenices and Bootes the Herdsman were climbing to the top of the sky, along with the beautiful golden K star Arcturus— the outriders of spring. He tightened his scarf and stuffed his hands in his pockets. His feet clanged on the catwalk. The fog had spread everywhere, drowning the lights of San Diego and Los Angeles, and had risen in a tide, lifting the domes of Palomar Mountain loose from their moorings on the world, to drift for a while above the shoals of mortality, below a heaven not exactly empty, but a long way from earth.

PART 2

The
Shoemaker
Comets

On a mountainside near Flagstaff, Arizona, a low house made of concrete blocks sat in a forest of ponderosa pines. It resembled a bomb shelter. A second house rested on top of it—a soaring structure with walls made of volcanic boulders and glass. At the dining room table inside the house made of boulders and glass, the astronomer and geologist Eugene M. Shoemaker examined a newspaper. He read aloud: "'Astronomers Locate Possible Distant Galaxy.'" He grinned and said, "What in the heck is this? What's this mean?" He put on a pair of half-glasses. "Cheezus, there's only about a hundred billion visible galaxies."

"I wonder which one they found," his wife, Carolyn Shoemaker, remarked dryly. She was clearing dinner dishes from the table. Night had fallen, and a steady rain poured down.

Gene dropped the newspaper on the table. He had a robust face tanned by years of prospecting for the remains of giant craters left by asteroids and comets that had struck the earth. He wore a thong tie with a clasp of Hopi silver in the shape of an eagle. He said to me, "Astronomers have essentially abandoned the solar system. In the nineteenth century the solar system was the object of central interest in astronomy. As their tools improved, astronomers focused their attention on what they called the larger questions. That's what the Hale Telescope was built for—to put enough horsepower in the optics to go after the structure of the universe." His wire-brush mustache conformed to his grin. He said, "So the geophysicists came along and adopted this orphan—the solar system."

A drumroll of October rain hit the house and reverberated along ponderosa rafters. Gene and Carolyn had lived with their children in the lower house until it had started to feel cramped, whereupon the roof had seemed a natural place to begin another house, especially one made of volcanic boulders. Eventually the children had grown up and moved out, and now Gene's mother lived in the lower house.

Gene said, "We are going to go after something new this month." He picked up a sheet of computer paper and unfolded it. It was titled "Known Trojans," and it contained a list of heroes from the Trojan War—Achilles, Patroclus, Hektor, Nestor, Priam. Each name was that of a minor planet in orbit around the sun, and each name was followed by a long set of numbers describing that planet's orbit. A minor planet is the same thing as an asteroid. The whole sheet of Known Trojans consisted mostly of columns of numbers. Lately, Gene said, he had been thinking about these Trojan planets, and he had begun to wonder if there might be a lot of *unknown* Trojans out there. He ran his finger across blocks of text composed of numbers. "Look at these orbital elements," he said. "You can just see from these numbers that the Trojan clouds are really enormous."

I could not tell anything at all from looking at the computer paper.

But when Gene Shoemaker looked at those strings of numbers, he could see in his mind's eye two immense, uncharted clouds of asteroids out by Jupiter. "These clouds," he said, "cover a heck of a lot of sky. The kind of sky we can explore with a little wide-field telescope."

"If the rain stops, Gene." Carolyn's voice came out of the kitchen, where she was washing the dishes.

A lugubrious sound hammered the roof. He looked up. "This is kind of discouraging," he said.

Carolyn came out of the kitchen. She said gently, "If it's raining in Flagstaff, Gene, then it's raining on Palomar Mountain."

What had sparked Gene's interest in the Trojan asteroids was the simple fact that Carolyn had recently discovered a new Trojan planet while she had been searching through some negatives. She had been looking for earth-approaching asteroids—stray asteroids closing in on the earth—but instead she had found this Trojan out by Jupiter. It was a big planet—a sooty ball about eighty miles across—by far the largest thing the Shoemakers had ever found. After they had photographed it enough to plot its orbit, they were entitled to give it a name. By long-standing tradition this type of planet must be named after a hero from the Trojan war. They studied a copy of *The Iliad*. In Gene's words, "The big names were all taken. We thought we were going to have to kind of scrape the barrel, you know. Get into the minor troops." Then they came across the name Paris. "For some reason, Paris had never been used. I don't know why. Paris was the guy who started the war." Paris was one of the sons of Priam, the king of Troy. Paris spent his youth mainly raising sheep. One day he stole Helen from King Menelaos, who was a Greek, and took her to Troy. Menelaos rallied his fellow Greeks to besiege Troy and get Helen back, and that was the beginning of the war between the Trojans and the Greeks.

There were two clouds of Trojans, one on either side of Jupiter and sharing Jupiter's orbit. Trojans were faint, slow-moving asteroids, and darker than anthracite coal, which was the reason why only forty Trojan planets had been found, whereas in the Main Asteroid Belt thousands of minor planets had been found. The Shoemakers had realized that the earth was about to make a relatively close approach to one of the two Trojan clouds of asteroids, accompanied by good viewing conditions. They had decided to devote some of their allotted dark time on the eighteen-inch Palomar Schmidt telescope to a search for Trojan asteroids.

The Trojan clouds had never been completely explored. Scattered pinpoints of light, barely resolvable on the photographic emulsion of a small telescope, Trojan planets were almost impossible to find. They fanned out for half a billion miles on either side of Jupiter. Nobody knew for sure how they had gotten there. Nobody knew for sure what they were made of—except that it was some dark substance.

In 1906, the German astronomer Max Wolf discovered an asteroid moving in the track of Jupiter's orbit, wobbling sixty degrees ahead of Jupiter, as if Jupiter were pushing it along. Wolf named the asteroid Achilles. Achilles had somehow wandered into a region of space where the gravitational fields of Jupiter and the sun formed a stable pocket of gravity, leaving the asteroid bobbing up and down in a limbo, unable to escape. In 1772, the French mathematician J.

L. Lagrange had predicted such a peculiarity in orbital systems. Lagrange had calculated that there would be dimples in gravity sixty degrees on either side of any body in orbit around another body. A stray object that happened to fall into one of these dimples would oscillate inside the dimple but never leave it again without a push.

Achilles was the first object found trapped in Jupiter's leading Lagrangian point. Then an asteroid was found in Jupiter's trailing Lagrangian point, traveling sixty degrees *behind* Jupiter, as if Jupiter were pulling it on a track. Soon it became clear that two swarms of minor planets bracket Jupiter, two clouds of black balls dancing on either side of the largest planet in the solar system. A tradition for naming these asteroids after heroes from the Trojan war became established. Asteroids that travel ahead of Jupiter are named for heroes from the Greek side of the war, while asteroids that trail Jupiter are named for heroes from the Trojan side—two seething armies of enemies, ruled by Jupiter. The two clouds are thus known as the Greeks and the Trojans, although astronomers usually refer to both clouds as simply "Trojans."

A search for asteroids required photographing points of light that brightened and dimmed like fireflies, as clouds and families of small planets swung around the sun, and the earth overtook and passed them. (The earth moves faster than most minor planets, because it orbits closer to the sun.) Now the earth was about to swing past the Greek cloud of asteroids. During a period of about three months the Greeks—the asteroids that traveled ahead of Jupiter—would slowly pass through the constellation Pisces. They would be visible against a dark, relatively star-free sky, far from the Milky Way, where dense clouds of background stars would otherwise camouflage faint asteroids. This month—October—marked the critical phase, because the core of the Greek cloud was now visible at the top of the sky around midnight. "So are you a crapshooter?" Gene asked me, glancing over his half-glasses. He said, "I think that we are looking at an opportunity to find a heck of a lot more of these guys."

The last time anybody had tried to take a census of the Trojan planets was twenty years earlier, when C. J. van Houten, a Dutch astronomer, had searched a few overlapping glass photographic

plates that had been taken of parts of the asteroid clouds. He had
startled some astronomers when he had declared that there might
be as many as nine hundred big undiscovered Trojans out there.
Startled or not, no astronomer had gotten around to checking van
Houten's theory.

Gene Shoemaker offered a more imperial guess as to the number
of Trojans. He said, "I think there could be two hundred thousand
Trojan planets bigger than a kilometer across, total, for both clouds.
We're talking about a heck of a lot of asteroids—a number roughly
the same order of magnitude as the *entire* Main Belt. I would add
that this is not the received wisdom about the Trojan planets." He
suspected that the Trojan clouds might extend far away from the
plane of the solar system, where nobody had ever looked systematically for Trojans before. "That's where we are hoping to hit pay
dirt," he said.

Asteroid clouds contain a range of debris—everything from dust
and sand through boulders to small worlds. If Carolyn, searching
films after the run, could discover a handful of big Trojans, that
would imply the existence of many small invisible objects. Big
Trojans, lurking in parts of the sky where they were not supposed
to, would betray a haze of unseen Trojans, swarming like no-see-
'ems on either side of Jupiter. Gene said, "These clouds could
actually be large shells of material." Carolyn would scan a pile of
negatives, looking not only for Trojans but also for earth-approach-
ers booming past the earth, because one of those guys could turn
up at any time, on any film. But if the weather failed them, they
would have to forget about Trojans for the time being. "This whole
thing is a crapshoot, anyway," Gene said, listening to the rain.

▪ ▪ ▪

The next evening, the Shoemakers loaded their Plymouth Fury for
the drive to Palomar Mountain. The Fury was a large golden green
boat with a smashed front fender. It appeared to have suffered
damage during the inexorable march of American science. As large
as the Fury was, the Shoemakers had a terrible time fitting their
gear into it. Carolyn owned a stereomicroscope that had to go to
Palomar Mountain, because she used it to discover comets and
asteroids. The microscope took up half the trunk, but the real

problem concerned a reporter traveling with the Shoemakers, who carried two backpacks and a duffel bag. Carolyn said, "What have you got in there?"

"Warm clothes," I told them.

"Eeyah," said Gene skeptically.

What I did not tell them was that I planned not to be the first reporter to freeze to death during a search for asteroids.

The Shoemakers preferred to drive to Palomar Mountain at night, because they could push the Fury across the Mojave Desert with less chance of meeting an individual whom Carolyn referred to as "John Law." Twenty miles outside Flagstaff, the Fury went *wahump*, and Gene said, "We're bottoming out on the shocks."

"Because you're driving too fast," Carolyn said.

They had wedged me into the backseat, beside my packs.

She turned around to look out the rear window at a pair of headlights gaining on us. "Is that John Law?" she asked me.

I turned around to look, and Gene turned around. "Where?" he asked. The Fury swayed and decelerated. A car whipped past us, going at least ninety miles an hour. False alarm—just another Arizonan in a hurry. Putting his foot back on the accelerator, Gene said, apropos of nothing in particular, "Carolyn had a good year last year. She got five comets."

She admitted that it was true. She said, "I got so that I expected to find a comet every time we went to the mountain. This year I haven't found any comets at all. I don't know what the problem is." After spending most of her adult life in the profession called housewife, Carolyn Spellmann Shoemaker had become an astronomer. Among other arts, she had learned the art of finding comets. She was a quiet, rather serious person, not given to mentioning her accomplishments, with a strong jawline and a hint of bronze in her skin each fall, which was the last of an outback tan acquired in western Australia. She and Gene spent their summers mapping giant ring structures in the landscape where comets and asteroids had hit the prehistoric earth. Her face gave people the impression that she possessed a secret compass that kept her moving along a private celestial meridian. She would tell people, "I think I was meant to be a hermit," by which she obviously implied that she was meant to be an old-fashioned astronomer. In the course of searching films for asteroids approaching the earth, Carolyn had begun to turn up comets. Some astronomers considered themselves lucky to find one comet in a lifetime—especially since comets are named for their discoverers. Carolyn had racked up so far a total of six Shoemaker comets, five of which she had discovered during one unbelievable eight-month roll in the year 1984. Before that, no astronomer had ever found five comets in eight months. Two of the Shoemaker comets are what are known as Jupiter family comets: Shoemaker 1 and Shoemaker 2 travel on short orbits around the sun that take them near Jupiter. They will gleam for perhaps ten thousand years before they go dark. Thus the name Shoemaker may last longer than marble or the gilded monuments of princes. The other four Shoemaker comets came through the solar system on long orbits and have now disappeared into deep space.

Among women, only Caroline Herschel, an Englishwoman who lived in the days of Jane Austen, had found more comets than Carolyn Shoemaker—eight—searching with a modest telescope that her brother, Sir William Herschel, had built for her. "I intend to beat Caroline," Carolyn remarked coolly. After that she was going to beat Mr. Honda, Mr. Bradfield, and Dr. Mrkos, three astronomers who were then tied for first place among living comet discoverers, with twelve comets each.

Each observing run on Palomar Mountain yielded a pile of photographic negatives, which Carolyn scanned in Flagstaff using her stereomicroscope. But the microscope went virtually everywhere Carolyn did. She brought it to the mountain in order to search films when Gene brought them fresh from the darkroom, since an earth-approaching asteroid could whip past the earth in a few days. When she searched for comets and asteroids, she inserted pairs of negatives into the microscope—photographs of star fields taken at intervals of forty minutes. An object in motion in the solar system would move enough during forty minutes to appear to jump out before her eyes, in stereo. Each pair of photographs contained around ten thousand stars or starlike objects. Most of them really were stars, yet the photographs were sprinkled with moving debris. Normal asteroids streamed in the same direction, like schooling fish. Abnormal ones, dangerous objects, things that could hit the earth, often went backward, against the normal flow of debris, or they slashed diagonally across the field of view, or they popped out, moving too fast. Carolyn had very sharp eyes, and she was always on the lookout for things that moved.

"Gene!" she said. "Is that car following us John Law?"

The Fury rocked as Gene turned around to look. "I hope not," he said.

"In Arizona," Carolyn remarked, "you can be jailed for speeding." The police had explained this to the Shoemakers before. She did not want a search for minor planets to end in jail.

The rain stopped and the clouds broke. A whitish pink star gleamed in the west, dead ahead. "Ah, Jupiter," Carolyn said. "A good sign." We switchbacked off the edge of the Colorado Plateau and down into the Mojave Desert. Carolyn propped a cassette player on the front seat and played a tape of Herb Alpert and the Tijuana Brass. The Milky Way arched overhead, and Jupiter dropped westward into basin and range. Jupiter is a big planet— the earth would fit easily inside Jupiter's Great Red Spot. Apart from the sun, Jupiter is the most influential piece of mass in the solar system. It seemed that Jupiter's gravity was warping Gene's foot to the gas pedal, pulling the Fury westward. In more than a metaphorical sense, Gene Shoemaker lived with his foot loaded on the accelerator. "I get cranked up over much more than I can finish," he said. "But I've resolved to spend my life, whatever time I have, in doing what's fun—"

"Nowadays you have to weed out what's really fun from what's just fun," his wife said.

"Yeah, I've got a bunch of irons in the fire, all intercon- nected—"

"If you could only stay off the committees—"

"Ha!" he said, meaning, "Fat chance."

Gene's scientific work in various fields had earned him some eleven medals and awards, which had piled up whimsically in boxes on top of the piano at home. He had served as a Principal Investigator with various NASA missions to the moon, and now he served as a member of the Voyager imaging team. If the field of impact geology—the study of what happens when a piece of rock or ice hits a planet—could be said to have a founder, it would be Gene Shoemaker. Caltech astronomers, weaned on the Hale Telescope, are quintessentially what are known as extragalactic types. To many of them the solar system is the deadest game in the house, offering little scientific challenge—nine balls of nonluminous matter whirling around a (pathetically) normal star, in addition to some gritty stuff, such as asteroids, moons, and comets: the two-dollar table in the grand casino of the sky. The following comments, which I heard in various places around Caltech at various times, give some idea of the attitude of many astronomers toward the solar system.

"If it's inside our galaxy, it isn't worth looking at."

"I simply cannot imagine looking for asteroids as a way to make a living."

"Planets are the slag heaps of the universe. The earth is a prime example of that. The only thing the earth is good for is to serve as a platform for a telescope. But we are going to have to get rid of this atmosphere. Then maybe we will see something interesting."

"I could care less if I found a comet. Unless it was going to hit the earth. Then I wouldn't want my name on it, anyway."

Gene Shoemaker offered an oblique reply to defamation of planets. "The solar system *is* an insignificant bunch of dust," he admitted. "It also happens to be where we live." Somewhere in his mind's eye, or maybe in his heart, Gene carried a peculiar vision of the solar system. It was not any solar system that I had ever heard of before. In schoolbooks the solar system is pictured as a series of flat, concentric circles centered on the sun, each circle representing the orbit of a planet. In Gene's mind the solar system was a spheroid. In Gene's mind the solar system was not at all the eternal, unvarying mechanism envisioned by Isaac Newton, but a carnival—a dynamic, evolving cloud of debris, filigreed with bands and shells of shrapnel, full of bits and pieces of material liable to be pumped into ellipses and loops and long, chaotic, wobbling orbits which carried drifting projectiles all over the place—minor planets that, every once in a while, would take a hook into a major planet, causing a major explosion. He said, "There's just a zoo of beasts out there, roaming the solar system. While it's tremendous fun discovering these little planets, the real fun is trying to find out what the heck they are and how they fit into the origin of the solar system." Curiosity, he had decided, was one of the two major forces that drove scientists, the other being a characteristically human wish to make a discovery for which one would be remembered after one was gone. "The trick," he said, "is to keep the bad ideas to a minimum." He grinned lopsidedly. "Which is not always possible." Then he swung the Fury into the breakdown lane and brought it to a halt. Time to switch drivers.

We piled out of the car onto a ribbon of highway that crossed a desert playa between mountain ranges. Not a pair of headlights was in view. Gene stood in the middle of the road and stretched. He leaned back and looked up. He said, "The Trojans would be dead overhead." His belt buckle glinted in the starlight—it was made of silver, in the shape of a many-rayed star. The sky glittered with lights, but the cloud of Trojan planets was invisible, twenty thousand times fainter than any star the naked eye could see. "Science," he suddenly remarked, "is not at all what you think it is."

I said, "It sounds like you and Carolyn plan to discover a new asteroid belt."

"Van Houten discovered it, in a sense—when he came up with that estimate of nine hundred Trojans. I just think it's a heck of a lot bigger than anyone else does."

A smell of damp creosote bushes filled the air. It had been raining in the desert. Not a good sign. "Heaven knows what it's doing on the mountain," Carolyn said. We climbed into the Fury, slammed the doors, and she floored the accelerator.

Gene began to talk about comets. One of the more intriguing shells of debris, to him, was the solar system's reservoir of comets. Beyond the outermost known planet, Pluto, there is a spherical shell of comets known as the Oort cloud, named after the Dutch astronomer Jan Oort, who demonstrated its existence. The Oort cloud contains a prodigious number of comets—anywhere from one trillion to one quadrillion comets (nobody can say how many) traveling in circular orbits around the sun, orbits that may average about a light-year from the sun. (If Pluto's orbit were the size of a dime, then a typical Oort comet would circle about ten yards out.) Comets are lumps of crumbly material, five or ten miles across, containing various kinds of ices, silica dust, and carbonaceous compounds. Comets are primordial pieces of the solar system, trash left over after the planets formed.

From out in the Oort cloud the sun would look like a bright star. Out there a typical comet travels slowly in relation to the sun—about three hundred miles per hour. Out there a comet can feel the gravity of stars other than the sun. The sun, Gene pointed out, was circling around the galaxy in the company of stars all around it (his vision of the galaxy ran parallel to his vision of the solar system: the galaxy was a collection of moving objects). All the stars in the sky were in motion around the galactic center, like traffic on a freeway. If a comet feels a gravitational pull from a passing star, the comet can, in some cases, be slowed down nearly to a halt—to a speed of around five or ten miles per hour in relation to the sun. Then it does what any object would do if suspended motionless over the sun. It falls toward the sun. By the time it reaches the inner solar system, the comet is falling at outrageous velocity. It takes a hairpin turn around the sun and heads back for the Oort cloud. Some comets actually hit the sun. Pondering comets dropping through the solar system, Gene Shoemaker wondered how often comets got trapped in the zone of planets. A comet could loop past Jupiter, for example; be slowed down by Jupiter's gravity; and wind up in an orbit near the sun. The zone of planets might be filled with invisible comets. The comets were invisible because they no longer sported tails. A comet began to steam as it neared the sun and the ices in it evaporated. The comet threw off dust as well. The result was the well-known display of a tail. If a comet was trapped in an orbit near the sun, then over time the ice in the comet nucleus might steam away.

One school of thought held that the ice in comets finally evaporated, leaving nothing but a whiff of dust behind. Gene disagreed. He suspected that a black nugget might remain—an extinct comet nucleus, maybe a mile across. "As the comet degasses, it grows a crust of dirt on it," he said, "like melting snow. You get the buildup of a lag deposit on the surface of the nucleus, probably some kind of polymerized hydrocarbon and rocky stuff, sort of like asphalt." The surface of an aging comet nucleus began to resemble a melting snowbank in the Bronx. The center stayed frozen, while the outer surface of the nucleus grew a skin of crud. As the crud thickened, the nucleus stopped throwing off dust. The tail vanished. Seen through a telescope, the comet now looked like a black asteroid. By definition it was now a minor planet on a chaotic orbit, a loose cannonball, rolling at large through the solar system.

A dead comet might wash up in a number of places. Its most probable fate would be to pass close to Jupiter and be whiplashed out of the solar system. Or perhaps it might get trapped inside the asteroid belt, to mix with the asteroids. Or it might hit Jupiter. Or—just possibly—it might end up colliding with the earth.

At three o'clock in the morning Carolyn swung the Fury into a truck stop in the Mojave Desert. We sat at a counter while a waitress in a checked pantsuit poured out three cups of coffee.

Had there been more comets at times in the past? Gene wondered out loud. Had there been comet showers? he wondered. What could cause a comet shower? He pondered these questions while he swigged coffee.

The waitress stood by the cash register with her arms folded, watching Gene. The coffee shop was otherwise empty.

Carolyn remarked, "Gene's wheels are always spinning."

"Yeah," he said, "and every now and then something kicks out." What would happen, he asked, if a star as big as the sun brushed near enough to the solar system to drill through the Oort comet cloud? He said, "I've calculated that we might have a close encounter with a big star maybe once every hundred million years. If a star the size of the sun went by slowly, it could give the Oort cloud one hell of a kick. That star could bore a hole through the Oort cloud. Comets would diffuse out of the cloud in all directions, which would increase the bombardment rate of the earth. We might be in the tail end of a comet shower right now."

The waitress came over. "Need 'nother coffee?"

"Sure," he said. "A comet coming in from the Oort cloud could do some damage if it hit."

The waitress filled our cups. She peered at Gene.

"These guys," he said, "travel at sixty-five kilometers per second, relative to the earth—that's triple the speed of the average near-earth asteroid." During a comet shower the earth might experience a series of bad random hits, like a chain of nuclear attacks. One result, in the past, might have been the wave of animal and plant extinctions at the end of the Cretaceous period, sixty-five million years ago, when something like half of the species on earth vanished, including the dinosaurs.

Gene went to the rest room, and the waitress chose that moment to come over with the check. She said quietly to Carolyn, "Have you all seen the Space Center up north of here?"

"You mean Edwards Air Force Base?" Carolyn asked.

"No. You know—where the starships landed."

"Oh?" Carolyn said.

"Where the aliens left those rock piles. You must've heard about that—since your husband is interested in that kind of thing. Those messages to outer space."

"That sounds interesting," said Carolyn.

"There's forces that keep the rocks together. Kids on motorbikes will knock the rocks away, you know? And the rocks will move back during the night. Nobody knows how it happens."

Carolyn paid the check.

The waitress added, "It might be some kind of magnetic force."

"We should go and have a look," Carolyn said.

"Don't miss it. Spacemen landed there. Have a nice night."

The Pacific Ocean sent a series of cloud fronts over Palomar Mountain, leaving the astronomers there feeling swindled. We stayed in a cottage on the hillside below the dome of the eighteen-inch telescope, staring out the windows over long breakfasts in the early afternoon. After breakfast we would hike or drive up to the dome, where Gene and Carolyn would tackle such work as they could find, while waiting for the weather to clear. The dome of the Little Eye, as astronomers sometimes called the eighteen-inch Schmidt telescope, was shaped like a bullet. The dome was eighteen feet across, offering more than a slight resemblance to a space capsule. It had two floors. On the lower floor were a tiny office, a darkroom, a supply closet, and a bathroom. On the upper floor stood the telescope. The entire dome would have almost fit inside the barrel of the Hale Telescope.

Gene worked in the darkroom. He mixed up tanks of darkroom chemicals. He unraveled a spool of Kodak IIa-D astronomical film, as big as a roll of paper towels, and chopped the film into six-inch disks, using a machine he called the Cookie Cutter. A boisterous racket came out of the darkroom. *Whomp.* Then a muffled "Damn." *Whomp.* The Shoemakers called these disks of film cookies. Gene filled three ammunition boxes with cookies. Ammo for the telescope. He carried the ammo boxes up to a laboratory at the Big Eye, where he injected them with nitrogen gas and baked them in an oven, to hypersensitize the films to faint light. "It's a black art," he said.

Carolyn set up her microscope in the dome's office and passed the time searching old films for asteroids. She always had a backlog of films. Her children marveled at her growing passion for minor planets, although perhaps they missed her a bit, because she spent so much time on Palomar Mountain. She listened to the radio while she worked, to a station that claimed it brought you absolutely the easiest listening in southern California. Every now and then she pulled apart the slats of a venetian blind to examine cloud streets and mares' tails sliding over the mountain. Life on Palomar goes catatonic under clouds. Down at the Monastery, astronomers sat around watching television, hoping for news of a break in the weather.

At the Little Eye, Gene spread out a sheaf of papers and wrote numbers in columns, planning the star fields they would photograph during that run; a geologist plotting a raid on heaven. This weather was beginning to get to him. He began to wonder if the Trojan planets would escape. He paced the office, looking over Carolyn's shoulder. One afternoon he suddenly vanished into the supply closet. Then his voice boomed: "Where did all these ants come from?" Tiny black ants plagued the eighteen-inch dome, and they had infested the supply closet. "Something really ought to be done about these ants," he said. But he had no idea what. He emerged with a jar of peanut butter and a spoon. He said to Carolyn, "Is this our peanut butter or has it been here awhile?"

"It's been here awhile."

"Want some?"

"No thanks, she said."

He pulled his half-glasses down his nose and inspected the peanut butter for ants. "Hm." He spooned a gob into his mouth. He swallowed it tentatively. Not immediately toxic, anyway.

"I've got something here that looks like a comet," she said.

"Oh, yeah?"

"A little fuzzy comet." She studied a loose-leaf binder containing recent notices of comets.

He sat down at the microscope. He couldn't find the comet. "Sometimes," he said, "I think I'm losing my—"

"Marbles?" she said, teasing him.

"Oh, yeah. It doesn't have a well-defined nucleus." They concluded it was only a galaxy.

Gene thought that he would call Bob Thicksten, the observatory superintendent. "Hi, Bob. Just wanted to see what our status is on the clocks. . . . Still slipping some? Uh-huh . . ."

"Rats," Carolyn said.

"We'll follow it carefully. What's the weather going to do? Grim? Ha, ha, well, the sky looks better than it has all day."

Gene's hopes for better weather proved to be unsupported by reason. A loud noise rattled on the metal dome. "What is that?" he cried, opening the door. A flash of lightning spilled into the office, followed by a rumble. "Aw, for crying out loud," he said. The ground was covered with hail.

This kind of thing went on for three days.

One afternoon Don Schneider showed up at the Monastery for breakfast. He poured some Rice Krispies into a bowl, juggled a hot Danish pastry out of the microwave, and said, "It's going to clear."

That jerked a few heads at the table.

"Have you heard a weather report, Don?"

"No, but Maarten Schmidt is arriving tonight."

"Oh, yeah. Maarten is supposed to have luck with weather."

"It isn't luck."

"What—does Maarten have a hot line to God?"

"No," Don said. "God has a hot line to Maarten Schmidt."

That evening, something in the gestures of the clouds tempted the Shoemakers to walk to the Hale dome, to get up on the catwalk for a view. They followed a road along the ridge, past scrub oaks and chokecherry bushes. Small flocks of birds flew in the wind. The air smelled of dead leaves and carried an edge of cold. A flicker of white and blue burst out of the underbrush, and a blue jay took off with a chokecherry in its beak. The Shoemakers circled the catwalk of the Hale dome, eyeing the clouds.

"Heaven knows what it will do," Carolyn said.

"A clear night's not impossible," Gene said hopefully.

As they so often did, they spent a moment admiring the Hale Telescope from a walkway inside the dome. "Anybody who isn't awed by that thing doesn't have a soul," Gene remarked. The Hale had something in common with the Hoover Dam, perhaps the naïveté of a world that still believed in its machines. The Hale Telescope embodied some of the longings of the twentieth century, and some of the terror. Gunn and Schneider were working in the cage at the base of the telescope, preparing for a quasar search. Something glittered in Gunn's hands: the pyramid of mirrored quartz that broke starlight into four beams.

The Shoemakers cooked some quick hamburgers in the cottage where we were staying, in case the weather cleared. We ate them quickly, watching the clouds through the window. At dusk, while we were sitting around drinking coffee, we noticed a tall figure walk by on the road, his hands in the pockets of his parka and his head down, lost in thought. Maarten Schmidt had arrived. Ten minutes later the clouds broke and vanished.

Gene and Carolyn stuffed the coffeepot into a paper bag, along with a pack of Oreos, and carried the bag out to the Fury. They drove up to the little dome and parked. While Gene worked in the office, Carolyn climbed to the upper floor of the dome and pulled a plastic sheet off the telescope (the dome leaked). She hit a button. With a screech bad enough to spall one's teeth, two curved doors on the dome pulled open. Hitting another button, she rotated the dome to the north, toward Cygnus, the Swan, a constellation that straddles the Milky Way. Twilight had eased off, and the black rift in Cygnus—a lane of dust in the Milky Way—was beginning to stand out. Hauling on a circle of hand holes at the telescope's base, she pointed it at the brightest star in Cygnus—Deneb—and looked at it through a guide telescope mounted on the barrel of the Schmidt. She fiddled with the Schmidt for a while, calibrating it. Not much bigger than a refrigerator, the telescope was coated with Dutch Boy battleship-gray paint. Rivets and dents on its tube suggested the hull of a submarine that had had some close calls with depth charges. Styled and built during the Depression, the Little Eye looked a bit like an aerodynamic pear, a design that its builders had evidently hoped would carry it smoothly through the winds of the future.

Gene came up the stairs. He put a pile of papers on a control desk next to the telescope. "The weight!" he said. The Schmidt had been slopping all over the sky (due to loose gears), and Jim Gunn had joked about hanging some lead on it to tighten down the gears. Gene had found a lead weight from a freight scale. He now lifted it from a shelf and hung it from the telescope on a loop of clear packing tape—Palomar Glue. He patted the Schmidt and said, "This oughta keep the gears tight." He grabbed the telescope, tipped it over until it pointed sideways, and snapped open two doors on its side. Carolyn handed him a film holder, which contained a circular piece of black-and-white photographic film—a cookie. He stuffed the film holder through the doors of the telescope, locked it in place, and snapped the doors shut. "Ready," he said.

Carolyn went to the control desk and read off the coordinates of the first exposure. "Right ascension twenty-two, thirty-two point zero."

Pulling by a hand hole in the base of the Little Eye, he slewed it across the sky, while dials on the wall told where the telescope pointed.

She said, "Declination plus fifteen, forty-seven." Her words smoked in the cold.

He slewed again. The telescope arrived at the edge of the Trojan cloud, now rising over the ridge to the east.

He sat on a stool, flicked off the lights in the dome. He peered into the eyepiece of the guide scope. He saw a set of crosshairs and a bright star—his guide star. He said, "What's the magnitude of this star, dear?"

"Six point four," she said.

The guide star lay near the crosshairs. "It's pointing to the right part of the sky, anyway," he said. He took up a control paddle. Hitting buttons on it, he tweaked the telescope until the crosshairs zeroed on the guide star. The telescope was centered on the first exposure. He said, "I'm ready."

"Five," she said, "four, three, two, one, open."

He reached up, pulled a lever, and two shutter leaves on the skyward end of the telescope opened like a pair of unclasping hands.

"Liftoff!" she said.

The telescope's automatic tracking drive would keep the crosshairs on the guide star while the sky moved and the telescope gathered light. Sometimes the guide star would go *oop*, slip off the crosshairs. Heaven doesn't jerk; telescopes do. He would work buttons on the paddle frantically to get the crosshairs back on the star before the photograph smeared.

The guide star suddenly kicked away from the crosshairs, and he said, "Bad hiccups!"

"We're looking east, Gene."

"It's horsing around. It's going all over the lot."

"That's a disaster, Gene."

"Ah, damn!" he said, and there was a sound like *zeee, zeee* while he punched buttons on the paddle. "This gear is worn," he said. "The telescope is oscillating back and forth." A flash of blue sparks danced around the telescope's mounting.

Carolyn said, "You might try putting the weight on the other side."

He switched on the lights. He put the weight on the other side of the telescope. Carolyn tugged the telescope to the coordinates of the next photograph while he studied a dial on the wall. "The clock could be slipping," he said. Bob Thicksten had warned him about that. "This is just incredible," he said.

Carolyn found the cassette player and tuned it to a radio station they liked, which was playing the Beach Boys, who were wishing that all girls could be California girls.

He turned out the lights and started another photograph. "This is awful!" he said, peering into the guide scope. "It's making huge excursions." He asked her for the fast-motion paddle. There were two control paddles for the eighteen-inch telescope: slow motion and fast motion. He needed both paddles to rein in this bucking bronco. "This is a two-fisted operation," he said over odd noises in the dark—*zip, zip, zip, click*—and more sparks fell from the telescope. He leaned back, sighed, and said, "You can see Jupiter."

Carolyn walked over to the dome slit to take in the view. "It amazes me," she said, "the way the weather will clear just like that. Looks like there's fog over San Diego. Good old fog." Surrounding ridges broke through the fog like the backs of whales.

"We'll have to get Maarten Schmidt to come up here more often," Gene said.

They traded places on the telescope, and Gene called off a set of coordinates to Carolyn. She hauled on the Schmidt, pitting herself against it; it weighed half a ton. She sat on a lift chair. Hitting buttons, she raised herself off the floor; the telescope was pointing laterally for this exposure and the eyepiece was hard to reach. "Horsefeathers!" she said. "I can't find my guide star. We are way off." They fiddled around the base of the telescope and finally got it pointed straight. She started an exposure, and the shutters flopped open.

"How's it treating you?" he asked.

"Not so bad now, Gene. Seems like there's always a shakedown."

The trick in sky photography is to get a telescope to track the stars closely while the earth turns. "When the telescope isn't tracking well," Carolyn explained, "the stars turn into seahorses and submarines in the photograph." The Schmidt's mounting looked like a tuning fork. The telescope's tube hung between the bars of the fork, and as the sky turned, the handle of the fork rotated to keep the tube pointed at the same place in the sky. Two dials on the wall told where the telescope pointed in right ascension (longitude) and declination (latitude) in a coordinate system known as the celestial sphere. Up until the time of Copernicus, astronomers had thought that the earth hung at the center of this celestial sphere and that the sphere revolved around the earth.

Each exposure lasted four minutes, and when an exposure was complete, the Shoemakers removed the film holder from the telescope, containing a single exposed film, and brought the film holder downstairs to the darkroom to be changed. They kept two film holders cycling, so that one was always inside the telescope gathering light. Next morning, Gene would develop the whole batch of films. As the night advanced, Jupiter set in the west, while the unseen Trojan cloud climbed to the top of the sky. The Shoemakers crossed and recrossed the cloud. They took strings of photographs, lapped at their edges like rows of fish scales. They changed places on the telescope once in a while. Every forty minutes they backtracked the telescope and rephotographed places where they had just been, to make stereo pairs. Anything that had moved during forty minutes would pop into stereo before Carolyn's eyes when she searched the films. The cold began to bite. Domes must be kept the same temperature as the outside air, since warm air inside a dome would flow out through the slit, rippling around the telescope and causing the stars to twinkle, which would hurt the seeing. A meteor crossed overhead, leaving a greenish, curly trail. Gene looked up from the control desk. "That was a beauty," he said.

"Darn," his wife said from under the telescope. "I always miss them."

"That's because you're looking through the telescope."

"It's not fair, Gene."

"Aw."

The little dome was crammed with scientific gear of obscure vintage. There were several stools of different sizes, a lift chair, and two mercury thermometers taped to the wall. There were clocks of all kinds, dead and alive. A sound of clicking relays from the control desk echoed through the dome. A generator hummed under the floor, and the stereo cassette played synthetic violins under the telescope. "Our son Patrick calls this 'music to go to the dentist by,'" Carolyn remarked. The wind drifted through the dome slit, bringing cold through as many layers of clothing as one could possibly wear. The stars were brilliant, intense, too close. Red numbers in a digital clock on the control desk flickered, dissecting time. I sat with my back against the dome and my feet curled up against the base of the telescope, listening to Gene and Carolyn talk quietly with each other. Their hoods were drawn, and they seemed artificially fat in too many clothes. They could have been astronauts. "Everyone seems to be getting divorced these days," Carolyn said. "I used to think we had a normal marriage. Now I'm not so sure." I cinched my hood and put on a pair of mittens. The dome slit seemed to be the window of a starship. I dozed off for a while, and when I woke up, there were strange stars in the slit— the sky had moved. We might have been lost in space. "Are you still with us?" Carolyn asked me. I remarked to her that it felt as though we were in free-fall.

"We are," she said.

She pulled a lever, and the shutters on the telescope's nose flopped open. "Actually," she said, "there was a time when I thought I would go into space. I won't now." A shower of sparks dropped from the telescope. "A lot of astronomy nowadays is electronic," she said. "Astronomers watch TV. For me it seems more real to be sitting under the open sky. I like to think that somewhere out there are these little bodies marching by."

Thoughtfully Gene said, "Those little bodies are generally one to three kilometers across."

"I think it would be nice to visit one," Carolyn said.

"It would be easier to reach some earth-crossing asteroids than to reach Mars," Gene said, standing at the control desk, monitoring the seconds fleeting past.

"What's my time, Gene?"

"Whoa," he said.

She reached up and pulled a lever. The telescope's shutters thudded down. She swung the telescope sideways. She reached into the telescope and unclamped the film holder in order to change the film.

The Main Asteroid Belt, Gene explained, is a collection of close to a million large pieces of rock and iron. Most of them float between the orbits of Mars and Jupiter. But every population has its rogues. When the Main Belt tossed off rogue asteroids, they could go into long, high, slanting orbits; they could approach the earth from any direction. "An earth-crossing asteroid," Gene said, "can appear essentially anywhere in the sky."

"How often do asteroids hit the earth?" I asked.

Gene said, without hesitation, "I've estimated that you'd get about ten major impacts every million years. Two-thirds of the asteroids land in the sea."

"What happens during an impact?"

Again without hesitation (he had it all figured out), he said, "If you took all the nuclear weapons in the arsenals of the United States and the Soviet Union—which I think is about twelve thousand megatons of warheads, based on the available public figures—and you put them all in one pile and detonated them, that's what we're talking about."

"I'm ready, Gene," Carolyn said.

He watched the clock. "Five, four, three, two, one," and then, lowering his voice, "open."

In 1932, Karl Reinmuth, an asteroid hunter in Heidelberg, Germany, discovered the first recognized earth-approaching asteroid, a lump of rock one mile across, tumbling past the earth in free-fall. It appeared on a glass photographic plate as a bright line among the stars—a near-earth object traveling quickly. He named it Apollo, after the Greek god who drove the chariot of the sun, because the asteroid's orbit took it near the sun. Apollo's orbit turned out to be unstable, chaotic, and astronomers lost track of it. Apollo popped up again in 1973, when it was rediscovered on another close approach to the earth. Apollo, as it happens, was also the god who fired invisible arrows at mortals to kill them instantaneously.

Five years after he sighted Apollo, Karl Reinmuth discovered another asteroid on a close approach to the earth—Hermes. Hermes boomed past us—missed us by about twice the distance from the earth to the moon, the closest approach by an asteroid that has yet been observed. That sounds like a fairly safe distance, but it is not. There are enough big earth-crossers roaming around to guarantee that once in a while one of them scores a bull's-eye.

The next discovery of an earth-crosser came in 1949, when a Palomar astronomer named Walter Baade was casually searching a glass plate taken on the forty-eight-inch Schmidt telescope on Palomar Mountain. Baade found a long, bright line tending in an abnormal direction—an asteroid that he named Icarus when he learned that its elongated orbit takes it past Mars, past the earth, past Venus, past Mercury, and close to the sun. Icarus is a cigar-shaped piece of rock a mile long, tinged brown and gray, apparently scorched by the heat from its visits near the sun.

Today less than half a dozen professional astronomers search systematically for wild asteroids traveling in unstable orbits near the earth. Not many astronomers are interested in objects close enough to the earth to actually hit it. We live under sniper fire, which can be witnessed every night—small meteors cutting the atmosphere. Farther out, mountains travel at hypervelocities through the void. A near miss, in astronomical terms, occurs often enough so that a diligent astronomer can earn glory photographing the event. On the Fourth of July, 1973, a former collaborator of Gene's named Eleanor Helin was working on the eighteen-inch Schmidt telescope on Palomar Mountain when she photographed an enormous, bright asteroid—roughly as big as the peak of Mount Everest—diving through the plane of the solar system and past the earth. During a twenty-minute exposure it made a line in a photograph that looked like a scratch. "It was going like a bat out of hell," Gene said. Shortly afterward it vanished. It was traveling on a cometary orbit, but unlike any comet, this object had no dusty tail. It was designated 1973 NA—not a name, a temporary label. 1973 NA will be back, but nobody knows where in the sky or when, because it is now a lost object.

Right after the new year in 1976, Eleanor Helin photographed an object loping alongside the earth, like a car edging up from behind in the fast lane of a freeway. It passed us. Then it moved directly across our lane. She named it Aten. Aten crosses our lane once every twenty years. There are two logical fates for Aten. Either Aten will one day pass so close to the earth that the earth will throw it into a new orbit. "Or," Gene said, "Aten's got a damned good chance of hitting the earth."

Gene and Carolyn Shoemaker have added quite a few earth-crossers to the list. One afternoon in May 1984, Carolyn was in the office at the dome of the Little Eye, searching transparencies of the night sky while Gene brought them fresh from the darkroom. She found a bright, slow-moving object. She telephoned the Minor Planet Center, in Cambridge, Massachusetts, and read the object's coordinates over the telephone. "The damned thing seemed to be accelerating," Gene would recall. A short while later a computer at the Minor Planet Center disgorged a solution to the object's orbit: it appeared to be accelerating because it was heading straight for the earth, like a headlight on an approaching train. This asteroid—1984 KD—zipped past, and it will return again and again, until it either hits the earth or is thrown into a new orbit. As of this writing, 192 earth-crossing asteroids have been discovered, most of them in recent years, and a fair proportion of them by Gene and Carolyn Shoemaker.

An asteroid is a relatively small piece of rock or iron, and the earth is a very large piece of rock and iron, and both are in free-fall around the sun. The wanderings of the earth-crossers take them through the orbit of the earth. Every now and then an unknown earth-crosser appears in the sky—a fast-moving point of light traveling at an average closing speed of fifteen kilometers per second, or thirty-four thousand miles per hour. For more than ten years Gene Shoemaker had been building statistics on earth-crossers, but the problem was that an Apollo object here, an Aten object there, added up to partial statistics. "Maarten Schmidt's got it easy with his quasars," he said, "because once you find a quasar, it doesn't move." Where did earth-crossers come from? How many were out there? A "close approach" seemed a rather large distance, except when measured against the huge size of the solar system, where watching a piece of rock close in on the earth at thirty-four thousand miles per hour seemed not unlike being stalled in an automobile on a railroad crossing at night and seeing a light down the tracks. At least you can abandon an automobile. As a statistician and thus a professional giver of odds, Gene Shoemaker saw, with a mathematical clarity, that in the fullness of geologic time what seemed impossible became inevitable: nature's version of nuclear war.

▪ ▪ ▪

One night I asked Carolyn, "How did you ever get interested in finding asteroids?"

"It kind of grows on you," she said. Her voice came out of the darkness under the telescope. "I wanted something to do. I get bored easily. I got tired of being a housewife, I guess. Gene's really a planetary geologist, not an astronomer. But he does a lot of things. His astronomy was something that really interested me. So I sort of eased into astronomy, part-time at first. In the beginning I didn't know if I could take staying up all night on the telescope with him. I worried about that. Especially in the winter, when it can get well below freezing in here and the nights last thirteen hours. I can't wear mittens, because you have to handle a lot of small metal parts when you reach inside the telescope."

"Close," Gene said.

She closed the telescope's shutters and continued. "But when I find a comet or a fast-moving object on a film, there's a surge of excitement." She rolled the telescope over on its side, accompanied by a whine of meshing gears. "I get a feeling I'm seeing something that no one else has ever seen before."

She thought it was time for a cookie break. She turned on the lights and went downstairs, returning in a moment with mugs of coffee and a bag of Oreos. "Would you like one?" she asked, offering me the bag. She had strong hands with rather short fingers.

Several Oreos and a mug of coffee helped to warm me up. I had dredged a pair of ski pants out of my packs, but nothing seemed to halt the cold. I had decided that the first maxim of survival on Palomar Mountain is that those who eat Oreos live to see the dawn.

"I think I would be a more doting grandmother," she said, "if I hadn't gotten so interested in asteroids." The pleasure of hunting minor planets had soaked up what she called a "restlessness." Being a mother had fulfilled her, she said, but she did not particularly enjoy being a grandmother. Her restlessness came from something deep, probably the American need to pile everything in a wagon and go looking for Eden. Her father, Leonard Spellmann, had tried silver mining in Colorado, without much success, so he went to New Mexico in 1920 to try homesteading. That might have worked out but for the fact that he met Hazel Arthur, a schoolteacher in Gallup, whom he married. Reluctant to put Hazel on a homestead, he ditched the homestead and moved into Gallup to open a men's clothing store. That might have been a success but for the fact that the Depression and his business partner cleaned him out. Around that time, Carolyn was born. Leonard picked up and moved the family to Oregon, where he sold insurance. That did not work out. He moved the family to Chico, California, where he sold real estate. There wasn't much of a market for rural California real estate in the Depression. She said, "Dad was basically a farmer; that was what he knew best."

"Close," Gene said.

She closed the shutters, extracted the exposed film holder, replaced it with a fresh one. A chain clinked. She started the next photograph.

In the early 1940s, her father bought a two-acre chicken farm in Chico. He killed and dressed the chickens, while her mother sold the birds to the local people by twos and threes. The family did not have much cash, and at first they found that beans better suited their dinner table than chickens. In the summers Chico became unbearably hot, and when the Spellmann family couldn't sleep for the heat, they would sit in the yard on blankets at night, and talk, and sing songs. Carolyn Spellmann was a dreamy girl and close to her family. When the singing stopped and the family dropped off to sleep on their blankets, she looked at the huge moon.

Her parents assumed that she and her brother, Richard, would go to college. Richard got into Caltech. The chicken farm could not handle more than one fancy tuition, so Carolyn went to Chico State College, where she could live at home and save on expenses. She eventually got a master's degree in education. While at Chico State she began hearing from Richard about his roommate at Caltech. Gene Shoemaker, meanwhile, began hearing from his roommate, Richard, about Richard's sister. By the time Carolyn met the touted Gene, they already liked each other. Then Gene left for graduate school in geology at Princeton University. They wrote letters to each other for a while, until Carolyn suddenly broke off writing.

Gene was upset. "What happened?" he asked in a letter. He kept writing to her, and to silence. She finally replied with an answer to the effect that "I thought that since you were at Princeton, you wouldn't be interested in me." She also mentioned to him that she and her parents were planning to drive around the national parks that summer. Gene wrote back asking if she might like to go camping with him on the Colorado Plateau. Carolyn's mother, Hazel, thought that the whole idea of her daughter camping in the desert with this Gene Shoemaker was just first-class. In fact, Hazel wanted to come along, too, as did Leonard. Then Leonard became sick and couldn't go, but Gene, Carolyn, and Hazel Spellmann drove around the Colorado Plateau in a car and told stories by firelight. At the end of the first week they went into Grand Junction, Colorado, to resupply. Gene and Carolyn, in a strategic move, left Hazel in a motel room one evening while they took the car out on the town. Gene suddenly pulled the car over to the curb and proposed to her. She replied, "No problem there, Gene."

"When I married Gene," Carolyn said, under the telescope, "the one thing that I hadn't quite figured on was what a workaholic he is."

"Ha, ha! Carolyn's pretty competitive herself," Gene said from the control desk.

"What's my time, Gene?"

"You have forty-five seconds."

She had tried teaching junior high school, hated it, and had gotten pregnant just in time to save herself from feral ninth-graders.

"I spent many years having a family," she said. "I enjoyed that. But I found I could get pretty restless. Then I sort of eased into minor planets. Once you start finding them, it's so much fun that you can't stop."

"Ten seconds . . ."

"Now I really cannot stop finding them."

"Five . . ."

"Sometimes I so much do wish I could stop finding them."

"Close." Gene said the word quietly, with an unmistakable satisfaction at another capture of photons.

▪ ▪ ▪

Carolyn Shoemaker was a watcher on the masthead, and she let out the first cry when an asteroid or comet came over the horizon. She could tell from an object's motion if it was something unusual. She would report such things immediately to an astronomer named Brian Marsden, who is the director of the Minor Planet Center in Cambridge, Massachusetts. Marsden would calculate the object's orbit on a computer. If the object turned out to be an earth-approaching asteroid, the Minor Planet Center would send out an international announcement of the discovery, thereby enabling other astronomers to perform detailed studies of the object as it passed by the earth. Carolyn had discovered (in addition to six comets) six earth-crossing or earth-approaching asteroids—1983 RB, 1984 KB, 1984 KD, 1985 TB, Nefertiti, and Mera ("a nymph, one of the lovers of Jupiter"). She had found these asteroids so recently that four of them had yet to be named. She had also discovered asteroids of various other classes (asteroids are often classified according to their types of orbits). She had discovered ten asteroids in the Hungaria class and fourteen in the Phocaea class. She had discovered the Trojan planet Paris, and also a giant asteroid out near Jupiter that she and Gene had named Caltech. She had also reported sightings of more than three hundred nameless Main Belt asteroids. She had, however, discovered and named quite a few Main Belt asteroids. Although these were not dramatic asteroids, like earth-crossers, they were perfectly decent minor planets, suitable for naming. One Christmas Eve, she and Gene put framed photographs of star fields under the Christmas tree.

Each photograph displayed a streak traced by a moving asteroid. Gene and Carolyn had named several asteroids for their children and other family members.

"I was just blown away," said their son-in-law, Fred Salazar, who received a minor planet permanently named Salazar, with the blessing of the International Astronomical Union, which oversees the naming of celestial objects. Fred's wife, Linda (who is Gene and Carolyn's youngest daughter), got the planet Linda Susan. Gene's mother got the planet Muriel. Gene and Carolyn's daughter Christy got Christy Carol, and their son, Patrick Gene, got Patrick Gene. Patrick's wife, Paula Kempchinsky, got the minor planet Kempchinsky. One of Paula's friends said to her, "My mom never gave me a planet for Christmas. All I get are pillowcases."

In 1801, Giuseppe Piazzi, a Sicilian astronomer, discovered a small planet in the empty region between Mars and Jupiter. Astronomers were not surprised. A wide gap between the orbits of Mars and Jupiter had intimated that that region might hold a planet. Piazzi named his planet Ceres Ferdinandea, in honor of King Ferdinand III of Sicily, which shocked other astronomers, because they felt that planets should be named after gods. So they shortened the name to Ceres. The next year, Heinrich Olbers discovered a second planet in the same area. He named it Pallas, for the goddess of wisdom. By 1807, Juno and Vesta had been found. Astronomers called these planets "asteroids," Greek for "starlike," because they were just points of light in a telescope. Many years went by without a new planet. In 1845, a postmaster named Hencke found the fifth asteroid, Astraea, thereby earning himself a pension from the king of Prussia. The land rush began. Before long, minor planets were turning up at the rate of five a year. One celebrated discoverer of asteroids was a German painter living in Paris named Goldschmidt, who had an apartment above the Café Procope. Goldschmidt aimed his telescopes out the apartment window at night. He bagged fourteen planets.

A tradition of naming asteroids for goddesses was soon firmly established, but by the time that astronomers got to Dynamene and Gerda the daughter of Gimer, they realized that they were running out of goddesses. They began naming planets after their wives, daughters, and female friends—Bertha, Edna, Rosa, Henrietta, Alice. During the 1890s, photography increased the discovery rate of asteroids to twenty a year, and the names veered away from the ladies. A Boston minister discovered Winchester, an exclusive suburb of Boston, in the Main Belt. An Austrian named a planet the Philagoria, for his Vienna club. Karl Reinmuth, the discoverer of Apollo and Hermes, also found Azalea, Geranium, Petunia, Chicago, California, and Granule (this last planet he named to honor the pathologist Edward Gall, who discovered Gall's Granule, a microscopic body inside white blood cells). A Russian discovered and named America. A Russian discovered and named Mark Twain. Russians also found Gogol, Chekhov, Jack London, and Rockwell Kent, not to mention Laputa, named after the floating island full of mad coprophilic scientists in *Gulliver's Travels*; but when they named a planet Karl Marx, that caused an international flap, although nobody seemed to mind when the Americans named a planet The NORC to honor a computer.

Clyde Tombaugh, of the Lowell Observatory in Flagstaff, discovered the planet Pluto in 1930. It is said that he found the ninth planet. This is an understatement—he discovered the 1,164th planet. Today, about 6,300 planets have numbers and known orbits. Another 65,800 planets have been seen once or twice, not often enough to have their orbits plotted with certainty and thus to be eligible for numbering. Newly numbered minor planets range in size anywhere from a few hundred yards across to a couple of miles across—or in the case of Trojans, fifty to eighty miles across. An explorer of the Main Belt named Edward Bowell is numbering a minor planet once every couple of weeks—using Clyde Tombaugh's Pluto telescope. Bowell said, "I am constantly up against the problem of what to name these things." He named a planet Barks, after his favorite comic-book illustrator, Carl Barks, who once sent Uncle Scrooge and the three little ducks on a voyage through the asteroid belt.

An asteroid is not nameable until its orbit is established, which requires three sightings on three separate trips around the sun. Then the asteroid receives a number and becomes eligible for naming. Asteroids are named according to the wish of the discoverer, provided that the name does not offend the International Astronomical Union. Up there (or down there, underfoot) drift the planets Kansas, Libya, Ohio, Pittsburghia, Atlantis, Utopia, Transylvania, and Paradise, the latter photographed for the first time by Schelte J. ("Bobby") Bus on the thirteenth of February 1977, on Palomar Mountain, and named after the town of Paradise, California, where his parents live. For as long as interest in the Main Belt lasts, the names Michelle, Davida, Douglas, Jerome, Dorothea, Anna, Iva, Diana, Mimi, Mildred, Dolores, Priscilla, Birgit, Oliver, and Iolanda will grace the lips of asteroid specialists. Dr. Paul Wild, of Switzerland, discovered the planet Rumpelstilz. Dr. Wild discovered Swissair, which he named for his favorite airline. He found Ragazza ("the Italian word for *girl*," as he explained in Minor Planet Circular No. 4,146, the international announcement of his discovery); Retsina ("in honor of the resined wine of Greece"); Cosícosí ("the Italian characterization of indifference"); and Bistro ("a small, cozy restaurant"). Three planets are named for Eva Peron: Evita, Descamisada, and Fanatica. Somewhere between Mars and Jupiter float the worlds of Fanny, Piccolo, Wu, Photographica, Requiem, O'Higgins, Lucifer, Tolkien, Echo, Zulu, d'Hotel, Fantasia, Limpopo, Valentine, Ultrajectum, Panacea, Geisha, Beethoven, Academia, Dudu, Felix, Bach, Chaucer, Einstein, Dali, Scabiosa, Nemo, and Mr. Spock.

How these pieces of rock came into being is an interesting story. About 4,550 million years ago, inside an arm of the Milky Way, a star exploded in a supernova. A supernova is the hecatomb of a star. There are at least two kinds of supernovas. One variety (Type II) begins with an aging giant star—a star at least eight times as massive as the sun. As the star grows old, it uses up its nuclear fuel while manufacturing elements. All of the lighter elements in the periodic table up to iron were probably created inside giant stars during late stages of nuclear burning, except for hydrogen, helium, and some lithium, which are primordial elements that were formed during the Big Bang. A giant star burns helium in its core during the last half million years of its life, creating carbon. It burns carbon for six hundred years, creating neon. It burns neon and oxygen to make sulfur, for about six months. The star develops onion layers of elements, all burning and fusing into heavier elements. At the center of the star grows an iron core, surrounded by a mantle of silicon. The iron cannot undergo nuclear fusion; it cannot burn. During the last day of the star's life, silicon at the boundary between the silicon mantle and the iron core burns rapidly. The silicon fuses into iron, which accumulates around the iron core. The iron core becomes too heavy to support itself. In a hundredth of a second the center of the core collapses. It implodes into a tiny ball of neutrons about the size of an asteroid—a neutron star. The onion-layered star is now a hollow shell. During the next three thousandths of a second the neutron star shrinks and rebounds. It snaps. That snap generates a shock wave that takes about a day to work its way up through the onion layers of the star, eventually blowing the star to kingdom come, generating a flash of light that can outshine a galaxy. The shock wave also ignites rapid nuclear synthesis in the fireball, yielding all the elements heavier than iron, such as silver, gold, and platinum, which join the star's other materials in a journey outward in a billowing bubble of gas and dust. A gold wedding ring begins with the death of a star. Everything in the human body except hydrogen comes from stars—carbon, oxygen, nitrogen in proteins, potassium and calcium in bone, iron in hemoglobin. Plato was right: humanity originated in the stars.

Shortly before 4,550 million years ago, a certain nameless star exploded in a supernova. The expanding shock bubble drove through a cloud of gas and dust inside an arm of the Milky Way and seeded the cloud with metals, while compressing parts of the cloud. At places along the leading edge of the shock wave, the cloud started to collapse under gravitation. In one place a knot of gas and dust flattened and began to spin. It happens that one of physical matter's more common habits on an astronomical scale is to collect into a rotating pancake of matter known as an accretion disk. The solar system began as an accretion disk. When the pressure and density at the center of the disk rose beyond a critical level, thermonuclear ignition occurred and the sun was born. The pancake flattened into a disk made of ice balls and rock balls. Called planetesimals, these were the ancestors of planets. The planetesimals collided and stuck together under mutual gravity, growing into planets. As the planetesimals orbited the sun, they separated into rings that probably looked like the rings around Saturn. Jupiter probably condensed first, from a thick ring, followed by the other planets, including the earth. When the planets fattened, their accretion rates slowed. They ate up their available planetesimals until only a few planetesimals were left.

Some planetesimals took their sweet time coming home. Gene Shoemaker's studies of the cratering rate of the moon showed that even about one billion years after the formation of the earth and the moon, late-arriving planetesimals still continued to pound into the moon, their impacts creating the lunar maria—seas of lava that welled up like blood from wounds on the moon's face. The earth must have suffered the same late heavy bombardment, and the earth must have once exhibited huge scars from impacts of late-arriving planetesimals, although weather erased these marks long ago. The late heavy bombardment has dwindled to almost nothing today. Almost. In Gene's words: "The last stage of accretion is still going on." The planets have never quite left off growing. The earth is now gaining about twenty tons a day, through a continual rain of dust from space. Every once in a while the earth gains two billion tons in one second.

Astronomers used to think that the asteroid belt might be the rubble of an exploded planet. Now they think it is the leftover material from a planet that never formed. Jupiter, the heaviest planet in the solar system, disturbed a ring of planetesimals in the region now occupied by the asteroids, preventing that ring from accreting into a planet. Jupiter's gravity raked those planetesimals, mixed them up, tossed them around. They could not stick together. Every time two planetesimals collided, they broke into fragments, and Jupiter pulled the fragments everywhere, causing more collisions and the production of more fragments. The asteroids are shattered bits of planetesimals that never congealed into a world; they are the bones of an accretion disk. Jupiter is still churning the Main Belt; accidents still happen. Most asteroids appear to be pieces of broken objects. Hammered by repeated impacts, asteroids are covered with a layer of dust and rubble, and some may even be piles of bashed fragments barely clinging together under their own gravity. Jupiter has already thrown most of the mass of the asteroid belt off into deep space. "If you took all the asteroids in the Main Belt and wadded them up into a ball," Gene said, "you would get something about a tenth of the mass of the moon. A spit in the bucket." Jupiter is still gradually grinding up the Main Belt and throwing its fragments away.

While most Belt asteroids are on stable orbits that do not come near the earth, it seems pretty clear to scientists who trace the entanglements of orbits that the Belt must be pumping asteroids into earth-crossing orbits. The Main Belt is itself gathered into rings, separated by clear lanes called the Kirkwood Gaps. Jupiter sweeps those lanes clean. Any fragment that falls by chance into a Kirkwood Gap enters a resonating dance with Jupiter, which can flip the asteroid away. Nothing can remain for long inside a Kirkwood Gap. Orbital specialists believe that the Kirkwood Gaps, and other unstable areas in and around the Main Belt, are a source of many earth-crossing asteroids. For example, two asteroids can collide in the Belt. A fragment can drift into a Kirkwood Gap. Jupiter can pull the fragment from the Kirkwood Gap and throw it into an orbit near Mars. If the asteroid happens to have a close encounter with Mars during the next few million years or so, then Mars can throw the asteroid inward toward the earth. As a result, the supply of earth-crossers is constantly being renewed. Jupiter drags asteroids from the Kirkwood Gaps and hands them to Mars, and Mars hands them to the earth. Saturn can also pull an asteroid from a Kirkwood Gap and throw it directly at the earth.

"A lot of astronomers call asteroids the vermin of the skies," Carolyn said.

Gene laughed, while his figure moved vaguely, outlined by a red reading lamp at the control desk.

"Gene and I," Carolyn went on, "regard *galaxies* as the vermin of the skies."

"There are far too damn many galaxies," Gene said. "Carolyn has nearly reported galaxies to the Minor Planet Center."

"They're confusing," she said. "The fainties can look like comets. I get so excited. Then I find out it's only a galaxy."

Gene's father, George Shoemaker, bought a farm in the 1930s along the North Platte River in Wyoming, where he raised navy beans. Beans were a lucrative crop during the Depression, and George's only problem was that his wife, Muriel, could not take farming beans. "My mother was gone like a shot," Gene said. "I guess if she had been able to stand it, I would be a farmer right now." Muriel Shoemaker left for Buffalo, New York, to teach school. Despite their differences over farming, George and Muriel stayed in love with each other and remained married. Gene would spend the winter in Buffalo and then take a train to Wyoming to spend the summer with Dad on the bean farm. Tiring of beans, his father went to Hollywood, where he eventually found work as a grip in a movie studio and where Muriel joined him again.

Gene went to high school in Los Angeles, where he became interested in radioactive minerals. He majored in geology at Caltech during the years following World War II. "Caltech," he said, "has always been a haven for space freaks." The tendency, he said, began with the Hale Telescope. He liked to stand in the viewing gallery of the Caltech optical shop and watch Marcus Brown's men, in white tennis shoes, work a polishing machine that traced Lissajous figures across the biggest piece of glass the world had ever seen. A few miles away, in Arroyo Seco, plumes of smoke occasionally erupted and a rumble shook the surrounding towns: Professor Theodore von Kármán and his students at the Jet Propulsion Laboratory were testing rocket motors. Then, during the summer of 1948, fresh out of Caltech, Gene found himself working for the United States Geological Survey, mapping uranium-bearing

formations in the Paradox Valley of westernmost Colorado. The
Geological Survey put him in a bunkhouse in a mining settlement.
He would drive into the town of Naturita each day for breakfast,
five miles on a dirt road through the Paradox Valley. One morning,
when he was pounding along in a Jeep on the way to breakfast, a
strange thought flooded over Gene. As he tells it, "I started thinking
about von Kármán and those rocket motors. I also knew what was
happening at the White Sands Proving Grounds. Wernher von
Braun was down there, firing off a bunch of captured German V-
2 rockets. All of a sudden I got this feeling in my bones. I said,
By God, they are going to build a rocket—*they are going to build
a rocket and take men to the moon with it!* What a thing! What an
unbelievable thing! To be the first man on the moon! And what
other person to explore the moon but a geologist? I decided right
there that when they took applications, I was going to be standing
at the head of the line." He saw a flaw in his plan, which was, as
he put it, "If you had told anybody in 1948 that you wanted to be
a geologist walking around on the moon, they would have considered you a prime candidate for the lunatic asylum." He swore an
oath to do whatever he could to get himself to the moon but to
keep his mouth shut about his ambition. At twenty years of age
in Paradox Valley, something terrible happened to Gene Shoemaker. He became a geologist obsessed with the sky.

A lunar geologist would have to know something about the holes
on the moon. In the late 1940s, prevailing opinion said that these
holes had been made by volcanoes. Gene taught himself explosive
volcanism. The earth's surface concealed many enormous, ringlike
features known as crypto-volcanic structures—believed to be the
remains of superexplosive volcanic eruptions. He studied crypto-
volcanoes. He also walked around Meteor Crater, a hole in the
ground nearly a mile across, outside Flagstaff. Despite its name,
"the majority of geologists," Gene said, "were equivocal—skeptical,
perhaps—that it was of impact origin." Some thought that Meteor
Crater might be a collapsed salt dome or a hole left by a volcanic
steam explosion. Not many professional geologists accepted a theory first proposed by Daniel Moreau Barringer in 1906, that a
nickel-iron meteorite had exploded on impact there. Gene set out
to make a geologic map of Meteor Crater for his Ph.D. thesis.

Barringer had drilled a series of holes in the floor of the crater, hoping to find a nickel-iron asteroid under the crater, which he never found. Gene examined Barringer's old core samples and discovered they contained a lot of shattered rock, which was full of microscopic droplets of quartz glass saturated with particles of meteoritic iron. Around the lip of the crater Gene found layers of sedimentary rock peeled back from the rim, "like the petals of a flower blossoming." He discovered that these layers of ejected rock had been deposited in reverse order. No volcano would lay down ejected debris in such an orderly fashion. For comparison he mapped craters formed by nuclear bombs in the Nevada desert— the Jangle U crater and the Teapot Ess. There he found thumb-sized blebs of shock-melted glasses blown into deeply shattered rocks, and sediments peeled back like flower petals from the lip of the crater, laid down in reverse order. The resemblance between nuclear and meteor craters seemed eerie to him. The evidence came to this: Meteor Crater had been made by an asteroid.

He stayed with the Geological Survey after receiving his degree. In 1960, Gene, Edward Chao, and Beth Madsen, all of the Geological Survey, discovered a natural mineral that they named coesite, found in the rocks of Meteor Crater. Coesite is a polymorph of silica that can form under shock—a wave of extreme pressure must rip through the rock, crushing the silica's molecular lattice into coesite. No known event at the surface of the earth other than the impact of a giant meteorite could do that. As Gene would later say, "We had discovered a fingerprint for impact." That brought him to Germany.

The Ries Basin is a circular depression seventeen miles across, north of the city of Augsburg, on the western border of Bavaria. Most geologists had assumed it to be an old volcano. "My German wasn't good," Gene said, "but the more I read about the Ries Basin, the more I became convinced it was an impact crater." He believed that with the coesite fingerprint test he could prove it. On July 27, 1960—six days after the discovery of coesite was first published— he and Carolyn drove into the Ries in a new Volkswagen bus. Around sunset they found a quarry—it belonged to a cement factory, and the workers had gone home—and climbed down inside. Gene broke a few pieces of rock with his hammer and looked at them in the fading light. At that moment, a scientific field, impact geology, came of age.

"The rock was shocked, melted, crushed," he said, "full of blobs of dark glass. I just knew instantly there was coesite in it." During the following days Gene and Carolyn explored the Ries. Gene walked through villages in a daze, with a rock hammer dangling from his hand. He found blasted, shocked rock everywhere, even cut into blocks and built into walls and houses. The Ries was a tremendous impact crater, populated with farms and towns. Near the center of the Ries they came across St. George's Church, in the town of Nördlingen, built of the Ries rock—a shattered and sintered granite speckled with oozy droplets of black glass. The medieval stonecutters had unknowingly put up a church to the God of the Apocalypse. Fifteen million years ago, during Miocene times, something had come in from space and exploded on impact. Crustal rocks had offered this object the resistance that a tub of lard would give to a concussion grenade. Rock blown from the lip of the Ries Basin had soared or slithered for miles across Bavaria. The Ries amounted to a Kepler or a Tycho—essentially a lunar crater in Europe.

That was the first proof that a giant impact crater existed on the earth. Gene's discovery opened the question of just how many impact craters the earth conceals, and it also opened the question of whether many of the so-called crypto-volcanic structures might actually be the eroded roots of impact craters. At last count, geologists have identified more than one hundred likely impact structures, including the sacred Lake Bosumtwi in Ghana; Lake Manicouagan in Quebec; dozens of eroded craters in the United States, including structures called Crooked Creek, Decaturville, Flynn Creek, Upheaval Dome, and the Manson Structure; the Serra de Cangalha in Brazil; the Rouchechouart in France; and Gosses Bluff in Australia. Gene thinks that perhaps as many as a thousand impact craters will eventually turn up, "provided we don't cover the earth with nuclear craters first." In 1960, when he walked into the Ries, the debate over whether the moon craters had been made by volcanoes still lingered; but if a big impact crater could be found on the earth, then those holes and rings in the moon would be impact craters too. Galileo had seen them the first time he looked through a telescope, but to show that they were made by asteroids and comets, and that the earth was pockmarked with similar rings, took three more centuries and Gene Shoemaker with a hammer.

He founded the United States Geological Survey's Branch of Astrogeology, now located in Flagstaff, dedicated to the geologic study of other worlds. He grew to prominence in the American space program, working first on the Ranger lunar probes, then as Principal Investigator for the imaging cameras on the Surveyor lunar lander, and finally as the Principal Investigator in charge of the geological fieldwork done during the Apollo manned lunar landings. But he never achieved escape velocity; he never left the earth. The adrenal glands on his kidneys progressively failed in 1962, killing forever his chances of going into space. "The irony of it," he said, "is that I chaired the committee that recommended the names of the first astronaut candidates to NASA." He would never forget the night launch of *Apollo 17*—the last of the manned lunar missions. He and Carolyn watched enthralled at Cape Canaveral as their friend and colleague from the Geological Survey, the geologist Harrison H. Schmitt, pulled away from the earth riding on a Saturn V rocket, which brightened and dimmed as it cut through cloud decks, a machine as tall as a thirty-story office building and already going at supersonic speed as it leaned to begin its roll downrange, while Gene studied, with the detachment of a scientist, the pain of the unfulfilled hope that had started that summer in 1948, in Paradox Valley, and which had delivered him to an open field in Florida, witnessing the launch of the first and last geologist to walk on the moon.

He would eventually leave the Apollo space program for other things, but he could not keep his eyes on the ground. Ever since Meteor Crater and the Ries, he had been wondering about rocks that fall from the sky. How many of them are out there? If you went looking for them with a telescope, what would you find? Could finding rocks in space give you a better estimate of how often the earth takes a hit? In 1972, when he began seriously thinking about a program to search for earth-crossing asteroids, the orbits of only three earth-crossers were accurately known: Icarus, Geographos, and Toro. Apollo had been lost. Hermes, the asteroid that had ambushed the earth in 1937, had also been lost (it still is). Astronomers had shown more interest in looking for exploding galaxies than loose cannonballs near the earth. Yet the bombardment of the earth was quite obviously a continuing natural process.

He began working with a geophysicist from Caltech's Jet Propulsion Laboratory, Eleanor Helin. Like Gene, Eleanor had begun to suspect that the number of earth-crossers might be enormous. She probed the Caltech archives looking for sightings of lost minor planets. She traveled to Germany to decipher the logbooks of dead astronomers—of Max Wolf and Karl Reinmuth—trying to recover the orbits of vanished earth-crossers. In 1973, Shoemaker and Helin founded the Palomar Planet-Crossing Asteroid Survey. She carried on the bulk of the telescopic work during the program's early years, spending long nights on the eighteen-inch and the forty-eight-inch Schmidt telescopes on Palomar. They endured gambler's luck. Right at the beginning, a huge Apollo object swept by, 1973 NA, now lost. "I said, 'Hot damn! We are on to something!'" Gene recalled. But then came a long dry spell with no discoveries. Then a burst of discoveries. Then another dry spell. "There were times when I was almost ready to give it up, but Eleanor Helin just would not quit."

She discovered Aten and Aristaeus, and codiscovered Ra-Shalom, all earth-crossers. She also discovered a large number of asteroids known as Amors, on unstable orbits near Mars—objects that could either hit Mars or be flipped into earth-crossing orbits in the future. Shoemaker and Helin defined three classes of earth-crossers. The Aten objects spend most of their time inside the earth's orbit. The Amor objects spend most of their time out around Mars, moving inward to brush the earth's orbit once in a while. The Apollo objects slash deeply back and forth. Gene has estimated that there are a total of about two thousand big Apollos, Atens, and Amors out there—asteroids able to collide with the earth now or at some time in the future—two thousand drunken mountains driving the freeways, most of which we have never seen. Smaller objects—the size of the Great Pyramid at Giza, for example—are exceedingly more numerous but exceedingly difficult to find. The odds are slim that something large might hit the earth during a human lifetime. From the human perspective, major impacts are rare. "Human civilization," Gene said, "is essentially instantaneous." From an astronomical perspective, hypervelocity planetoids slam into the earth rather frequently. We live in an asteroid swarm.

Shoemaker and Helin eventually decided to divide their program. Helin founded the International Near-Earth Asteroid Survey—a program to coordinate sightings of earth-crossers all over the world. Shoemaker opted for a small but intense program on the Little Eye. Having neither the time nor the patience to scan films for asteroids, he needed an assistant.

It had not taken Gene long to propose marriage to Carolyn, but after they had been married, it had taken Gene about two years to work up his courage to tell her that he intended to go to the moon. She was frightened. She wondered if her husband was unstable. On second thought, that did not seem too bad, considering that she had always wanted to go to the moon herself, since those summer nights in Chico during her childhood; and so their hope became a mutual affair. She would explain that during the 1960s, "I thought that travel to the moon would become so common that even someone like me would be able to go." Neither of them had made it into space, but at least there was no harm in going up on a mountain each moonless part of the month between September and May, to photograph a dome of jewels out of reach and to swear at a telescope.

The eighteen-inch Palomar Schmidt telescope was a wide-field telescope; it yielded virtually a panoramic view. In one snapshot the Little Eye could photograph an area of sky larger than the bowl of the Big Dipper. The Little Eye contained two pieces of glass— a twenty-six-inch mirror and an eighteen-inch corrector glass at the nose. (Schmidt telescopes are rated in size according to the diameter of the corrector glass, not the mirror.) The Little Eye was one of the smallest professional telescopes in use anywhere in the world, and it bulldozed the sky. The Hale Telescope, on the other hand, drilled thin holes into lookback time. Even using electronic cameras, the Hale Telescope would require more than a human lifetime to make an overlapping mosaic of pictures of the northern sky. The Little Eye surveyed the entire northern sky more than once a year. The Big Eye had never caught an unknown asteroid cruising near the earth. "The eighteen-inch," Gene said, "is the fastest gun in the west." He felt, however, that small telescopes did not attract government funding or the attention of private donors. The Shoemakers' search for asteroids that could hit the earth was costing six thousand dollars a year. That was apart from Gene's salary, which the Geological Survey covered when he was on the mountain. Much of the six grand went to pay for rolls of film. "The cookies," Gene said, "are two dollars a throw." The Shoemakers had been trying to raise some grant money to cover a salary for Carolyn, without luck. The Geological Survey would have been happy to cover her salary, but federal rules on nepotism prevented that. She had no choice but to value her time at nothing. She found earth-crossers for free. Gene said, "We're getting too much of a bang for these bucks." Dents on the Little Eye's tube suggested that it had been getting more bangs than bucks for a long time.

Bernhard Schmidt, the inventor of the Schmidt telescope, was born in 1879 on an island off the coast of Estonia called Nais Saar, which means The Island of Women—a whaleback of fields and forests in the Baltic Sea, five miles long, with a lighthouse at one end. The islanders dressed in the old costume and took their Lutheran religion seriously. Bernhard was a troublesome boy with a scientific bent who took bombs seriously, and he began designing and building them for fun. One Sunday morning, at the age of eleven, he skipped church and went into the fields to set off a pipe bomb of his own design. The material inside the pipe detonated unexpectedly, yielding a blast that must have rattled the windows of every church on Nais Saar. The blast also took off the sleeve of his Sunday suit, which unluckily had contained his right arm, which was also gone. He washed the bloody stump in a brook and waited in the woods for the churches to let out, and then ran home, afraid he would be punished for drenching his suit with blood.

After losing his arm, Bernhard Schmidt turned to the science of light. He began to grind lenses as a hobby, and when he grew up, he left the Island of Women. Around the turn of the century Schmidt landed in the town of Mittweida, in Germany. There he set up a workshop in an abandoned bowling alley, where he ground mirrors for amateur astronomers and lived on the small amounts of money that he got for his polished glass.

The difficulty in making astronomical mirrors lies in grinding a concave surface that will gather starlight and bring it into sharp focus. An astronomical mirror is a light scoop designed to pour a large quantity of light into a tiny area, and the wider and deeper the scoop, the more light it can deliver rapidly to the film. In the language of optics, a mirror that can gather faint light quickly is said to be a "fast" mirror. The use of a fast mirror allows for brief exposures on film, thereby speeding up the astronomer's work. A fast mirror is a steeply curved mirror. The curve known as a paraboloid is particularly good at gathering faint light, except that parabolic mirrors suffer from an unavoidable optical defect: only a tiny area at the center of a photograph taken with a parabolic mirror contains stars in good focus. Toward the edge of the photograph, the stars smear into commas. Astronomers avoid this defect by restricting the size of the film to a small area in the center of the field of view. In the Hale Telescope, for example, the area of good focus is about three-eighths of an inch square at the focal plane— an area the size of one's little fingernail. Parabolic mirrors are terrible for searching large areas of sky.

Bernhard Schmidt became a master at shaping fast parabolic mirrors with his left hand. He kept himself going on brandy, cigars, coffee, and sweet cakes. With a fishtail of cigar stuck to his lip, Schmidt paced over his mirrors in the bowling alley all night, touching them a little with a polishing tool held in his left hand, keeping his empty right sleeve pinned up to prevent it from dragging on the glass. To get into the bowling alley he had to walk through the Lindengarten Restaurant. He told the proprietress, Frau Bretschneider, "Put out a good bottle of brandy for me, and if as I go by I pour myself some, then I'll make a mark on the beer mat." Schmidt kept his telescopes in an empty lot across the street from the Lindengarten Restaurant. On cold winter nights, when he was observing stars and simultaneously polishing glass, he kept rushing back and forth through the Lindengarten Restaurant, each time having a shot of brandy. The brandy got Schmidt coming and going, and by the end of a night the beer mat was covered with marks. "We used to make out very well with him," Frau Bretschneider remembered.

Schmidt's telescopes appeared to be screwed together out of scrap lumber and vegetable crates. One of them, a solar telescope, had a heliostat mirror that tracked the sun by means of a dripping water clock. He was shy and aloof; he never married. There is no evidence that he liked women. He may have been homosexual, except that there is no evidence that he liked anybody, much. He was a frank pacifist. "Only one man alone is worth anything," Schmidt once said. "Put two men together and they quarrel. A hundred of them make a rabble, and if there are a thousand or more, they'll start a war." So they did, in 1914. The German police began to watch him. They had no idea what was going on with the mirrors and the water clock, but they knew he was a pacifist, and since he was an Estonian, they concluded that he was an Estonian traitor flashing signals to Russian aircraft. They put him in a prison camp, where he suffered terribly. When the war ended, he returned to his bowling alley to continue grinding lenses.

In the years following the First World War, Schmidt's work came to the attention of professional astronomers in Germany. Professor Richard Schorr, director of the Hamburg Observatory, at the risk of getting himself into trouble with the police, managed to extract Schmidt from his bowling alley and install him in a single-men's dormitory at the observatory's branch in the town of Bergedorf, outside Hamburg. There Schmidt lived for the rest of his life, a volunteer member of the staff and a chronic alcoholic, making mirrors for the observatory whenever he felt like it. They called him *der Optiker* B. Schmidt—The Optician B. Schmidt.

Professor Schorr gave him a basement room for a workshop, but *der Optiker* appeared to spend much of his time wandering around Bergedorf while talking to himself, apparently drunk, and puffing a tilted cigar under a brown felt hat pulled so low over his eyes that people feared he would catch the brim on fire. *Der Optiker* was terribly secretive. He allowed almost nobody to enter his underground vault. Opticians do not mix easily with humanity. As Mel Johnson, the optician who did some of the final figuring on the Hale mirror, explained to me, "All that scratching and dirtiness with people, it just kills you off." The optician fears that people will feel the glass, and who knows where their hands have been. People will scratch their scalps, putting dust into the air, which can get between a polishing tool and the glass. One time, unbeknownst to Schmidt, a visitor picked up a polishing tool and took a few strokes with it on one of Schmidt's mirrors, removing about a tenth of a millionth of an inch of glass. When Schmidt tested the glass, he saw skid marks where the visitor's tool had touched down. "Somebody has been fooling around with this!" he screamed. Professor Schorr always felt that Schmidt's real skill as an optician lay at least as much in his eyes as in his hand, for when Schmidt looked at glass through his testing instruments, he could tell instantly where it deviated from a perfect optical surface.

Schmidt had such respect for glass that he dressed himself in a formal morning suit (a man's long-tailed wedding suit) before he entered his polishing chamber. He hung a straw boater hat on a nail, revealing cropped, graying hair. He stood over a disk of glass and walked slowly in circles around it, watching the glass with a stern expression on his face, stroking the glass now and again with a small figuring tool faced with pitch, and kept his cigar stub unlit or let it burn in an ashtray to prevent ashes from dropping on the glass. His left hand looked like the hand of Michelangelo's Moses— gnarled, monolithic, nourished with blood vessels. He knew the superiority of his hand to any polishing machine. "My hand is more sensitive than the finest gauge," he said. His mirrors rested on bizarre contraptions made of crates and boards and ropes and pulleys. He refused to tell the world how he shaped his glass. "If I were to write it down," he said, "it would so shock the astronomers and the opticians that I'd probably never get another order to construct anything."

In the summer of 1929, an eclipse of the sun happened in the Pacific Ocean, and the observatory dispatched Bernhard Schmidt and a young astronomer named Walter Baade to the Philippines to photograph it. They left Hamburg in February on a steamship and did not return until September. The eclipse was blanketed by clouds, ruining some of the observations, which seems a pity for Schmidt and Baade after such a long journey. During the trip Baade snapped a photograph of Schmidt. In it Schmidt has no arms at all. Schmidt had hidden his one arm behind his back, because he was holding a bottle and he didn't want it to appear in the photograph. The expedition's main event, at least for the history of astronomy, took place on the steamship somewhere in the Indian Ocean.

Walter Baade never wrote much, but he left some stories with friends. What happened went something like this. It was evening, as Baade later recalled. While Baade did not describe the setting, I think it is likely that the two astronomers were standing at the rail of their ship, watching a dark tropical sea slide past. The air would have been so clear that the stars contacted the horizon. Under the spell of the Clouds of Magellan, *der Optiker* would have steadied himself at the rail with his hand and remarked, around a fresh-lit cigar, that he had thought up a design for a telescope.

Walter Baade listened.

This telescope, Schmidt said, would be able to photograph a huge area of sky in a single exposure, and the star images it rendered would be needle-sharp from edge to edge on the film. It would be a very fast telescope.

Baade sensed that Schmidt had been thinking about this for a while.

First, Schmidt said, he would grind a mirror into a deep, hollow, spherical curve. That was easy—any two-fisted optician could achieve a spheroid. The reason that nobody used spherical reflectors in telescopes was that they produced horrendous distortions across the entire photograph, making such mirrors useless for astronomy. But what if one could install a correcting glass on the snout of the telescope, to reshape the light falling on the mirror? The correcting glass, Schmidt said, would have a toroidal ripple in its surface, an undulation so subtle that the glass would look to the untrained eye like a piece of flat window glass. But it would not be flat. It would introduce subtle distortions in the light. When the light bounced off the spherical mirror and hit a piece of film in the center of the telescope, it would be in complete focus from edge to edge of the photograph.

Baade was stunned. The potential of such a telescope must have been immediately obvious to Baade: it could be used to search the sky for moving objects. Baade told Schmidt that he must grind one of these corrector glasses as soon as possible.

Noch nicht! Schmidt said. Not yet! First, he said, he must think up a method to polish his corrector glass. He would never, never, do a sloppy job on a piece of glass. Never! The praxis, the technique, he said, must be *sehr elegant*.

When the expedition returned to Germany, Baade and the observatory's director, Professor Schorr, began to press Schmidt to make his corrector glass. Schmidt answered them by walking calmly and aimlessly around Bergedorf, drunk, threatening to catch his hat on fire with his smoldering cigar. "Above all, he valued his independence," recalled Baade. Abruptly Schmidt took off for the Island of Women. At some point—it is satisfying to think that it might have happened while Schmidt was tramping around those fields where he blew off his arm—he dreamed up another of his bizarre machines for polishing glass. He returned to Bergedorf. To Baade he remarked that he needed to know something about the bending characteristics of a thin sheet of glass. Baade gave him a handbook of physics. Schmidt studied the book and then donned his wedding suit and shut himself in his vault. He wanted nobody to see what he did.

When Schmidt had been in his vault for thirty-six hours, Baade began to worry about him. Baade finally went downstairs and found Schmidt unconscious near a wafer-thin, fourteen-inch disk of glass. "On waking," Baade said, *der Optiker* "accepted cigars but declined coffee and sandwiches," because, Schmidt said, he still had twelve hours of polishing left to go. Schmidt smoked a couple of cigars and shooed Baade out. The work climaxed in a frenzy of polishing.

The finished corrector glass was so thin that a pull with a pair of hands could have snapped it. As for his "elegant method," Schmidt had placed the glass on a pan, like putting a lid on a cook pot, so that it formed a seal. He then pumped the air out of the pan while the glass bent downward and was sucked into the pan. He polished his glass flat, released the vacuum, and the glass popped up into a rippled shape. *Sehr elegant*!

By the summer of 1930, Schmidt had built a telescope to house his glass. One muggy Sunday afternoon he took Baade up to an attic window in one of the buildings at the Bergedorf observatory for first light. Ordinarily it is not possible to look through a Schmidt telescope by eye (the light comes to a focus in the center of the tube), but for this occasion Schmidt had evidently rigged up some kind of a prism to bring the light to an eyepiece. Schmidt pointed the telescope across a wide meadow, toward the Neuer Friedhof cemetery.

Baade squinted into the eyepiece. Baade noted a clarity of colors.

He noted that the edges of the leaves on the trees in the cemetery were razor-sharp.

"Can you read the names on the tombstones?" Schmidt asked.

"Yes," Baade said, in awe. "But I can see only one thing: the optics are absolutely marvelous." Baade then asked Schmidt just how big a corrector glass could be ground.

Forty-eight inches across, said Schmidt. Not any larger.

Then they put film in the telescope and photographed a tombstone. The letters of the name were clear.

That summer and autumn, Schmidt photographed the sky. Light passed through the glass, down onto the mirror, and bounced halfway back up the tube until it came to rest on the film—a simple and powerful imaging system. The films were circular and spangled from edge to edge with veldts of the Milky Way. One winter night he and Baade pointed the telescope horizontally and made an exposure of a windmill two miles away. The sails came out crisply. When they looked at the photograph through a magnifying loupe, they could count individual twigs on distant trees. As a matter of fact, it was a moonless night; the twigs had been illuminated by starlight.

The Bergedorf observatory was immensely proud of its new telescope. But when *der Optiker* offered to build Schmidt telescopes for other observatories, nobody wanted one. He lowered his price to a pathetic sum. No other observatory in Europe would touch his camera, the fastest telescope on earth. *Der Optiker* grew loud and resentful in the taverns of Bergedorf. When he had taken a lot of brandy and had become what he called drunk *"auf Achse"* ("drunk to the axle-shaft"), he would order a round for the house and proclaim, "The whole world is going to hear of Schmidt someday!"

The 1930s in Germany were not a good time or place for a one-armed Estonian with a criminal record for pacifism. Schmidt smelled another war coming, which made him extremely angry, and so he contrived a method to escape the war: he began to thoroughly rinse his bloodstream with expensive cognac—an elegant praxis—and it worked. In the winter of 1935, "Death took the polishing tool from his hands," as one of Schmidt's colleagues wrote, referring to Schmidt's "hands" in the plural, seeming to forget that Schmidt only had one hand. They buried him in the same cemetery that he and Walter Baade had photographed while testing the first Schmidt telescope. It is the Neuer Friedhof cemetery in Bergedorf, near Hamburg. If you happen to visit that graveyard, I recommend that you turn through the right-hand door at the main entrance of the cemetery and follow a ring path until you reach a magnolia. You will now be standing among the very tombs from which Baade and Schmidt took first light. There you will find *der Optiker*'s own grave, which is a dark headstone engraved only with his name, a star, and the words PER ASPERA AD ASTRA.

▪ ▪ ▪

In 1931, Walter Baade joined the staff of the Mount Wilson Observatory in Pasadena, bringing with him the photograph of the tombstone that he and Schmidt had taken, as well as some photographs of the night sky. These greatly impressed George Ellery Hale and everyone who saw them, including a Caltech physicist by the name of Fritz Zwicky. Soon afterward, Zwicky collaborated with Walter Baade in a great discovery—they found that stars can explode with extreme violence. They used the word *supernova* to describe the explosion. Fritz Zwicky longed to watch a supernova go off. A supernova is a rare event, and Zwicky realized that he would be unlikely to see a supernova anytime soon inside the Milky Way, but he figured that with a wide-field telescope he could monitor a large number of galaxies and perhaps thereby raise his chances of seeing the gleam of a supernova. A Schmidt telescope would be perfect for a supernova search. Construction of the two-hundred-inch telescope had barely begun on Palomar Mountain, but Zwicky began badgering opticians and engineers at the Mount Wilson Observatory and at Caltech to build him a telescope. The result was the eighteen-inch Schmidt telescope, or Little Eye, now used by the Shoemakers to search for comets and asteroids. Russell Porter, who styled some of the details on the Hale dome, designed the eighteen-inch telescope's aerodynamic lines.

First light hit the mirror of the Little Eye in 1936, when Fritz Zwicky began photographing swarms of galaxies in Virgo, hoping to catch an exploding star. He had luck. He found supernovas popping off in galaxies all over the sky. Zwicky's eighteen-inch Schmidt was the first telescope on Palomar Mountain, and the only one there for the next twelve years, until the Hale Telescope went into operation.

In 1947, the forty-eight-inch Palomar Schmidt Telescope saw first light. This telescope was Walter Baade's jewel. He had supervised its construction. It had a corrector glass of exactly the largest size that Bernhard Schmidt had said would work, that summer afternoon in Bergedorf. When Baade's telescope went into operation, Zwicky's little Schmidt dropped into obscurity inside a thicket of carrasco oaks, and Zwicky began to feel ignored by his colleagues and especially by the press. He began to feel that the other astronomers, Baade in particular, did not want him to use the Hale Telescope. Zwicky probably did not help his requests for time on the Hale when one night, when he was working at the forty-eight-inch Schmidt, he ordered a night assistant to hurl a series of cherry bombs out of the dome, hoping that the explosions would improve the seeing. The seeing did not improve, but the noise and flashes, which made it sound as if Zwicky had started a war near his telescope, did not encourage the other astronomers to let him use the Big Eye.

Zwicky began to remark loudly that *he* had built the first Schmidt telescope. Baade reminded Zwicky that Bernhard Schmidt had. Zwicky became irritated with his former collaborator, Walter Baade. Zwicky began referring to Baade as "the Nazi." That was a cruel joke. Baade was an excitable, rather delicate man, with pointed ears and a bow tie. He limped badly—one of his legs was a good deal shorter than the other—and he stuttered. He was no Nazi. His hands trembled from nervousness, giving some of his colleagues the impression that he was about to come apart at the seams. Yet somehow whenever Baade took the guide paddle of a telescope in his hands, all the shaking stopped, as if Baade were transfixed at the sight of his guide star in the way that a deer can be jacked in a flashlight beam; and then Baade took masterful photographs of star fields that were as fine as powdered talc.

Zwicky had his own ideas about how to photograph a galaxy: he thought that photographic emulsions should be mixed with explosive chemicals. That way you could point the telescope at a galaxy, open the shutter, and you would hear a little frying sound and a pop inside the telescope when the light hit the film, causing it to explode—now that was what you called a fast film. Zwicky was a maximal space freak. In the early days of rocketry, Zwicky put an explosive charge on the nose of a German V-2 rocket, and when the rocket had reached the top of its trajectory, Zwicky triggered the charge, which fired a scrap of metal off into deep space. Zwicky was proud of that because he, Fritz Zwicky, had sent the first human artifact into escape velocity from the earth. Zwicky held some fifty patents, including one for an underwater ramjet that he called the hydrobomb. He regarded most of the other Palomar astronomers as fools, and Walter Baade as a cretin. Zwicky, who had been born in Bulgaria but raised in Switzerland, believed that he was superior to the others not only mentally but physically. He tried to demonstrate this by doing one-armed push-ups on the floor of the Caltech Athenaeum, a posh dining room on the Caltech campus. There the faculty ate filet mignon and reasoned with one another on scientific topics—and on at least one occasion collectively stopped their forks midway to their mouths while they watched Fritz Zwicky flop to the floor like a bull seal and in a grunty, roaring Swisso-Bulgaric accent, challenge anyone to beat him at one-armed push-ups. The Palomar astronomers could not get rid of Zwicky—he had tenure at Caltech—but they consulted psychiatrists to see if he was edging into psychosis, and the outlook must not have been favorable, because Walter Baade grew physically afraid of Zwicky.

It was not difficult to feel afraid of Fritz Zwicky. He had a glowering flat face, pale blue eyes, and a savage sense of humor. He would swear torrentially at night assistants, using scientific terms laced with obscenities. He referred to Baade and the others as spherical bastards—"They are spherical," he said, "because they are bastards every way I look at them."

Zwicky was, in fact, a true genius, and one of the greatest minds at Caltech, although his personality kept him in deep trouble with his colleagues, some of whom clearly hated him. He made many discoveries. His most prophetic happened in 1933. Studying the motions of galaxies in the Coma cluster, which is a cluster of galaxies fairly near the Milky Way, Zwicky realized that these galaxies were moving abnormally quickly around their cluster. The galaxies were moving so fast that the whole cluster should fly apart. But obviously the cluster was not flying apart. He concluded that some kind of powerful, unseen gravitational force was holding the cluster together. He did not know what it was, so he called it the missing mass. For many years astronomers tried not to think about Zwicky's missing-mass problem (one tried not to think about Zwicky at all), until recently, when astronomers finally could not deny that the universe really does contain large amounts of unseen mass. They now call it the dark matter. And it is a very big problem. They do not know what the stuff is, although they do know that it makes up as much as 99 percent of the universe. In other words, astronomers do not know what most of the universe is made of; and Fritz Zwicky told them so. His discovery of the missing mass is probably the most important problem in modern astronomy. After all, it would be nice to know what the universe is largely made of.

Zwicky used to say, "Only Galileo and I really knew how to use a small telescope." According to one astronomer who knew Zwicky, Zwicky seemed to fill up more space than he actually occupied, as if he himself contained missing mass; and Zwicky fully occupied the little dome, slamming the eighteen-inch telescope around, allegedly putting dents on its tube while he looked for exploding stars. Walter Baade began to wonder what would happen if Zwicky went mad. What if he burst out of the little dome and came around to the Hale Telescope one night looking for Walter Baade—and not with any telescope, either? Hands shaking, Baade whispered to colleagues that he believed Zwicky was going to murder him.

Rumors that Fritz Zwicky was on the verge of murdering Walter Baade got around. During dinners at the Monastery, Baade and Zwicky sat at opposite ends of the table, where they did not speak to each other and hardly to anyone else, although Zwicky's pale eyes flicked in Baade's direction, giving some of the diners the need to take an antacid. An astronomer named Milton Humason was often present during those dinners. Humason had begun his career as a janitor and a mule driver on Mount Wilson, and had been promoted to astronomer. Humason had collaborated with

Edwin Hubble on the discovery of the redshift of the galaxies and thus of the expansion of the universe, one of the most important scientific discoveries of the century. Humason was a short, humble man who wore a Chicago-style felt hat and a heavy coat that generally came equipped with a pint of Jack Daniel's somewhere in the coat, for extra protection against substellar cold. Milton Humason was regarded as the kindliest of the astronomers on Palomar Mountain, but Humason had been around mules long enough to know when to draw the line. One night at dinner, Zwicky gave the final proof that he had gone mad. Zwicky said in a loud voice that a rocket ought to be fired at the moon in order to recover moon rocks for study.

"Aww, Fritz!" Humason rumbled. "Leave the goddam moon to the lovers!"

Walter Baade retired to Germany, where he died of natural causes. Fritz Zwicky continued to work at Caltech into the 1970s. He ended up in a basement office in the Robinson building on the Caltech campus, down among the astronomy graduate students. Watching them pass by his office, every once in a while Zwicky would roar out, "Who in the hell are you?" He died in 1974. The asteroids named Zwicky and Baade drift in the Main Belt today and are not likely to collide with each other anytime soon.

■ ■ ■

The Shoemakers decided to take a break for coffee and apples. They walked over to the dome slit and looked at the sky, which had developed into one of those autumn nights when the stars grow needle-sharp, immanent.

"Andromeda galaxy's up," Gene said. "Right at the top of the sky."

"I can never find it, Gene."

They both leaned out of the dome slit, silhouetted against the Milky Way. "Look at the Horn of Plenty," he said.

"Got it."

"See the mouth of the Horn?"

"Sure."

"Two stars near the end. Now go up from those two stars—"

"Which way?" she asked.

"That way. See a diffuse patch? Andromeda galaxy's bigger than the full moon."

"Oh! It looks different, Gene."

"What? We've been looking at it in the guide scope."

"Well, it looks different from here."

Gene turned around and propped his elbows on the ledge of the dome and bit his apple. "What a beauty of a night," he said.

"Gene, we need a pipeline to God," she said, "to tell us where the asteroids are."

▪ ▪ ▪

Maarten Schmidt, who liked to philosophize on the catwalk of the Hale dome, remarked there on one occasion, "I cannot believe that people become astronomers in order to earn money, or to become well respected. In so many observatories they work alone. In the freezing, freezing cold. Everywhere one finds these people studying just one star. Just one star! What it means, I don't know." He paused. "It means they are all slightly mad." He fell silent. Suddenly a meteor cut through the sky.

"Nice!" Maarten whirled around.

The meteor moved rapidly until it burst in a flash that left a spot before the eyes. A piece trailed off and faded out. Maarten put his hand up to his ear, listening. "We might hear a sonic boom," he said. A minute went by. "Ja, well," he said, "the time is getting on." He descended to the data room, where Jim Gunn and Don Schneider were watching galaxies traverse the screens. "We saw a very bright, beautiful meteor," Maarten said.

"Fantastic," said Jim. "Did you hear anything?"

"No. We listened."

They passed around a glass jar of Oreos and discussed the meteor. Maarten said, "The sound would have taken five minutes to reach us. It exploded. A little piece shot off." He spread out his fingers. "Like this—pop!"

They had seen the destruction of an earth-crossing minor planet the size of a golf ball. When it hit the atmosphere, the resulting flash had been visible for upward of three hundred miles. "The Shoemakers," Jim Gunn remarked, "are into one of the few areas of astronomy that may have a real impact."

■ ■ ■

Sometime before the year 1664, a shower of meteorites fell into Milan, Italy, killing some sheep and a friar. A curious doctor who opened the dead friar found that a small stone had shot the friar through the femur, shattering the bone and causing his death. In 1856, a bright light passed over the ship *Joshua Bates*, covering it with black-glass dust. On the morning of June 30, 1908, a fireball passed over Siberia, making "even the light of the sun appear dark," according to a witness. In the minutes that followed, something terrible happened in a deserted swamp in the valley of the Middle Tunguska River. A brilliant mushroom cloud boiled up into the stratosphere. Fifty miles from ground zero, a shock wave picked up a tent full of Evenkian nomads and tossed it like a purse through the air. A man sitting on a porch at the Vanovara trading station, seventy miles from ground zero, experienced a bath of radiant heat, followed by a concussion wave that plucked him from the porch and threw him ten feet through the air, knocking him senseless. The blast incinerated and flattened forests over hundreds of square miles, and the roar broke windows and crockery up to six hundred miles away. A pressure wave traveled twice around the earth. The following night, the sky glowed so brightly over Europe that a person in London could read a newspaper outdoors at midnight; and the cause of it all was the impact of a comet or an Apollo asteroid about two hundred feet across.

One afternoon in 1912, at Holbrook, Arizona, the section boss of the Santa Fe Railroad and his family were eating dinner. They heard a "terrific crash." One of the boys ran outside. He said, "It's raining rocks out here!" Father went out. As they looked, the entire plain for a mile eastward filled with puffs, reminding them of bullets "kicking up dust." Fourteen thousand meteorites landed.

In Johnstown, Colorado, on July 6, 1924, a funeral was taking place in the graveyard behind the little church at Elwell, when there came a sound like machine-gun fire, and then a *thug* as a meteorite hit the road where the funeral procession had just passed. Mr. Clingenpeel, the undertaker, dug it up.

On April 28, 1927, in Aba, Japan, Mrs. Kuriyama's five-year-old daughter was playing in the garden when she cried out. She had been struck in the head by a meteorite the size of a mung bean, which her mother found resting on her neckband. This stone now sits in a museum in Japan. It is called the Aba.

One bright day in 1931, Mr. Foster of Eaton, Colorado, was leaning on his hoe in his garden. Something hummed past his ear and whopped into the soil. He pulled out a coppery nugget shaped like his pinky. It was the Eaton. He sold it to a collector for five dollars.

In Benld, Illinois, Mrs. Carl Crum was working in her backyard on the morning of September 29, 1938, when she heard what sounded like an airplane going into a power dive, followed by a sound of breaking boards coming from her neighbor Mr. Ed McCain's garage. "Just one of those things," she thought.

That afternoon, Mr. Ed McCain thought he would "go uptown" in his Pontiac coupe. He went into his garage. He opened his car door. There was a hole in the seat. He called to his neighbor, Mr. Carl Crum: "Come over here, Carl, and see what the rats have done to my seat cushion."

Mr. Crum came over and studied the damage.

Mr. McCain went on, "I knew the rats were getting thick around here, but I never supposed they could do that much damage."

"Ed, no rats ever made that hole."

Then they saw a hole in the roof of the car. They backed the car out of the garage. They lifted up the seat cushion, and there, tangled in the seat springs, they found the Benld.

An Air Force pilot flying at high altitude over Alabama on November 30, 1954, saw a bright light, like a falling star, heading in the general direction of the town of Sylacauga. Meanwhile in Sylacauga, across the street from the Comet Drive-In Theater, Mrs. E. Hulitt Hodges had just fallen asleep on her couch, when a loud noise woke her up. She jumped to her feet. At first she thought the gas heater had exploded. Then she felt a pain in her side. The falling star had destroyed her roof, bounced off her radio, and bruised Mrs. E. Hulitt Hodges viciously over the hip. Now, lying insolently on her rug, was the Sylacauga—eight and a half pounds of hypersthene stone, fresh from out near Mars.

After the nuclear test ban treaty of 1963, the United States Air Force established a secret global network of air-pressure sensors in order to detect any clandestine aboveground nuclear blasts. Within a year or two the network had detected a number of tremendous air shocks, including one over the South Atlantic that registered half a megaton. Either somebody was cheating or—Air Force scientists realized—meteoroids exploding in the upper atmosphere were releasing the energy of a nuclear bomb.

April 8, 1971. Wethersfield, Connecticut, dawn. A witness saw a bright bomb bursting over the town. An hour later Mr. Paul Cassarino went into his living room and found plaster on the floor. He looked up and discovered the Wethersfield One embedded in the ceiling.

On August 10, 1972, something came in from space over Utah. For two minutes it moved northward over Idaho and Montana, going at least Mach 20. It may have skipped off the atmosphere over Canada, or, Gene Shoemaker suspects, it may have coasted to a relatively gentle impact somewhere in the Canadian forest. Mr. James Baker, vacationing at Jackson Lake, Wyoming, took an extraordinary photograph of it. In the picture his wife is standing on a dock. She is obviously startled. She looks toward the Grand Tetons, where, high over the peaks, a fireball is leaving a ruler-straight smoke trail through the upper atmosphere. This was almost certainly an Apollo object somewhat bigger than a diesel locomotive, overtaking the earth from behind. Had it come in at a steeper angle, Mr. Baker might have photographed a two-kiloton mushroom cloud boiling abaft the Tetons.

Wethersfield, Connecticut, again. One evening in 1982, Mr. and Mrs. Robert Donoghue were watching *M*A*S*H* in the breezeway of their home when they heard "a sound like a truck coming through the front door." The Wethersfield Two entered their living room and battered the walls as it ricocheted around, overturning furniture. The Donoghues left in a hurry. Firemen found the stone under the living room table.

Perth, Australia, September 30, 1984. At Binningup Beach two sunbathers heard a whistle and a thump. A meteorite had missed them by twelve feet and burrowed into the sand. Three months later, in Claxton, Georgia, Mr. Don Richardson was just stepping out of his trailer when a whine that reminded him of mortar fire in

Vietnam made him flinch and he saw a hostile incoming chondrite destroy Mrs. Carutha Barnard's mailbox.

Gene and Carolyn Shoemaker are cosmic weather forecasters. The chance of falling stones, they warn, is 100 percent. Gene estimates that about once a year, somewhere over the earth, a meteoroid vaporizes in the upper atmosphere during a Hiroshima-force "event," as he likes to call such happenings. Since water covers two thirds of the earth, many airbursts happen unwitnessed over the sea. Every twenty-five years, perchance, there is an event that nudges a megaton—the power of a hydrogen bomb. An event like the Siberian Tunguska blast will occur roughly once every three centuries. Gene gives 5 to 20 percent odds that an impact twice as powerful as Tunguska will occur during the next seventy-five years. He figures upon only a 1½ percent chance that a 670-megaton blast will happen in the next seventy-five years. That could probably snuff Belgium. "I'm not losing any sleep," Gene said. "The odds are much better that we will do it to ourselves first."

What if Carolyn actually found an object heading for the earth? "I might not notice it in the films," Carolyn said from the control desk. "If it were heading right at us, it wouldn't appear to move. It would look like a star. Not until it was on top of us, when it would suddenly appear to move very fast." But by then, she added, it might be too late to tell anyone.

Gene dragged the shortest stool under the telescope and con-torted himself in order to fit on it. "Naw, this isn't going to work," he said. He kicked the stool aside and sat on the floor. "This is a terrible angle," he said, peering into the guide scope. "Suppose," he said, "that a twelve-megaton fireball went off over a politically unstable region. Suppose that happened over Pakistan, and suppose Pakistan had the bomb. The heat, the light, getting knocked on your fanny by the airwave—a large number of people would swear they had been nuked. The political leadership might say, 'Oh, those S.O.B.'s! They've nuked us!' And respond with a real nuclear attack on someone else."

Over geologic time, sooner or later, there would come to pass what Gene liked to call a "major event." That was to distinguish it from an "event." Major events happened roughly once every hundred thousand years, and many happened in the sea, of course. If the Tunguska horror had been made by an object the size of a building, then a responsible scientist had to consider what might happen if a projectile the size of the summit cone of the Matterhorn clobbered the earth. Gene knew what would happen. A Matterhorn coming in vertically, he said, would punch through the atmosphere in one second. He said, "A bow wave in front of it opens a hole in the atmosphere, and the atmosphere burns, making nitrogen oxides." He turned on the lights and studied a dial on the wall. He said, "I think this clock is slipping again." He pointed a flashlight at the base of the telescope and fiddled with some metal parts. After he had gotten things under control he explained that when a giant asteroid moving at nine miles per second hits the ground, it explodes. "I'm ready," he said.

Carolyn counted down. He started an exposure looking to the west, where the Trojans were now setting, because dawn was approaching. He said that when the asteroid lands, a hypersonic compressional wave rips through the asteroid, transforming it into a liquid mass that tries to splash but turns into a fireball. The rocks at ground zero compress to one-third their normal size, and a tremendous flash of light floods out of ground zero. "The radiant heat," Gene said, "would set buildings on fire sixty miles away." If the asteroid hit the sea, it would start tsunamis (tidal waves) cresting all over the planet. If a major event occurred in the sea off Long Island, then—never mind Manhattan—forget New Jersey. Forget Boston, Washington, Miami, Lisbon, and Dakar, for the spreading tsunamis would obliterate many cities sitting on the rim of the Atlantic Ocean.

The impact would squeeze out a ring of debris—a conical sheet of crushed, molten, and vaporized rock known as the cone of ejecta. The leading edge of the cone of ejecta expands at hypersonic velocity upward into the atmosphere, like a blossoming flower. It superheats the atmosphere into a bubble of gas mixed with molten and vaporized glass. The bubble bursts through the top of the atmosphere into outer space, and the glass keeps going. Thumb-sized pieces of glass might soar on ballistic suborbital trajectories halfway around the earth to reenter the atmosphere at a concentration zone opposite the point of impact, causing fire storms of molten glass to rain down over a region the size of Australia. The blast might inject enough dust into the atmosphere to blot the sun over the entire planet, perhaps causing a temporary winter. Nitrogen oxides (the burned atmosphere) would turn into nitric acid rains. Gene suspected that during the last million years a spike of powerful impacts had occurred—a pulse of perhaps as many as thirty major events, including as many as twelve continental impacts. This suggested the possibility that our species had come of age during a mild comet shower. It seemed that *Homo sapiens* had already survived a dozen natural versions of nuclear war—with the important difference that impact holocausts do not leave radioactive fallout in their wakes.

There are asteroids much larger than the Matterhorn traveling on earth-crossing orbits. Sisyphus and Hephaistos, two earth-crossers, are each about six miles in diameter. Either could hit the earth. If either did, the result would be what Gene called a "global catastrophe" to distinguish it from a mere "major event." The first strong evidence that life on earth might have endured a global catastrophe came in 1980, when Luis and Walter Alvarez and coworkers analyzed an unusual layer of grayish and reddish claystone found near the medieval town of Gubbio, Italy. This claystone dates from about sixty-five million years ago. It is thin—less than an inch thick— but it marks a sharp boundary between beds of rocks below it, containing Cretaceous fossils from the age of dinosaurs, and rocks above it, containing younger Tertiary fossils from the age of mammals. Now known as the K-T boundary layer, it contains abnormally large amounts of rare metals, such as iridium, which are found in much higher abundances in meteorites and comets than in crustal rocks on the earth. The K-T layer has been collected from more than seventy locations everywhere on earth, including the ocean basins. It resembles a coating of paint that was once laid down over the entire planet. The K-T layer also contains microscopic spherules of minerals that may once have been molten glass, shocked mineral grains, and carbon soot. Sixty-five million years ago the sky misted glass, forests on several continents burned, and the planet grew a skin of clay.

The Alvarez group proposed that the impact of a large Apollo asteroid had thrown up enough dust to cause an impact winter, during which global temperatures dropped for months or years, halting photosynthesis in plants. At roughly the same time at least half of the species on earth vanished, both plants and animals. "If you put that thin layer of clay up in the atmosphere as dust," Gene remarked, "it wouldn't let in any more light than a slab of wet clay half an inch thick. The only light at the surface of the earth would have been from luminescent organisms and from fires." Darkness had triggered the death of much single-celled life in the sea, dynamiting the pyramid of life at its base, causing mass extinctions to ripple upward.

Gene had teamed up with Cesare Emiliani and Eric Kraus to try an idea: that a single projectile, about six miles across, had dropped into the Pacific Ocean. Erupting with the force of a thousand nuclear wars, a shock bubble blew thousands of cubic miles of ocean, atmosphere, and crustal rock into outer space, leaving a hole filling with lava erupting from the earth's mantle. In one hour a blanket of dust, spreading in free-fall through space, cloaked the earth. The sea returned over the crater as a tidal wave several miles high, which boiled when it hit the lava. The crater might still be preserved in a deep ocean basin—you might find rings on the sea floor. "But we just haven't explored the oceans that well," Gene said. The rock and seawater thrown into space came down as mud rains or mud snows over the whole planet, mixed with nitric acid, but a considerable amount of water vapor stayed aloft. Water vapor traps heat from the sun. Global temperatures, after plunging during the time of darkness, bounced upward when the skies cleared. A greenhouse effect had cooked the dinosaurs and turned shallow seas into hot tubs. Small mammals, able to shed heat by virtue of their small bodies, had managed to maintain breeding populations. "That's one of the ideas you could put into the pot," Gene said.

Gene's style as a scientist was to consider the possibilities. When he encountered a promising idea, he felt obliged to try it. By no means had he ruled out the possibility of an intense comet shower. Perhaps a star had passed through the Oort cloud of comets. The Oort cloud would have gone berserk, throwing comets in all directions, some of which would have poured through the solar system, hitting the earth once in a while, causing stepped or staggered mass extinctions.

Global catastrophes seem to have occurred throughout geologic time. Paleontologists have found evidence for a great dying among the soft-bodied creatures of Precambian times. They have found evidence for two mass extinctions *during* the age of dinosaurs—the first happened at the end of the Triassic period (it may have been caused by the comet or asteroid that made a ringlike puncture in Quebec known as Lake Manicouagan), and the second extinction happened at the end of the Jurassic period. The final extinction of the dinosaurs may have occurred in multiple waves, as if a comet shower had occurred. Or perhaps a giant asteroid had shattered in the Main Belt, injecting a rain of fragments into earth-crossing orbits. During the age of mammals there had been two more horrors, one of them a staggered burst of mass extinctions at the end of the Eocene period.

Biological evolution, it seemed, consisted of periods of stability followed by abrupt developments. After a mass extinction, surviving organisms would branch out, evolving into new forms of life. The evolution of the mammals had been one such explosion, perhaps. The extinction of the dinosaurs might have been heralded by a star shining during the day—a star passing through the Oort cloud—and by an unusual number of comets hanging in the dawns and sunsets, year after year, as the Oort cloud dropped rogue comets, disturbed by the passing star. But if an asteroid had hit the earth, there would have been no warning at all, except the presence of Jupiter in the sky, serene and delicious, and, as always, raking the asteroid belt, flinging asteroids at the earth. If not for Jupiter, perhaps, the dinosaurs might have had time to become slim creatures examining quasars with mirrors, while today we would be balls of fur with no brains and saucer eyes, chewing insects and howling at the night.

▪ ▪ ▪

"We'll put you to work tonight," Carolyn said to me.

"What are we doing?" asked Gene from under the telescope.

"I'm going to teach him how to change film."

"Good."

She said to me, "Watch Gene's time and do the countdown for him when his time is up. This is a four-minute exposure."

The red numbers on the digital clock hurried along, dividing time into tenths of a second.

"God damn you!" Gene said to the telescope, not to me, and the telescope answered him with a flash of sparks. I heard *zeee*, *zeee* while he got the crosshairs back on his star. He closed the shutters and rotated the telescope. After some clanking in the dark Gene gave me a film holder containing an exposed cookie of film. I gave him the fresh film holder. Next I had to run to the control desk.

"While Gene is slewing the telescope," Carolyn said, "you read off the coordinates of his next exposure. Hit the buttons to rotate the dome. Hand him his control paddle. Write down the time, the temperature, the name of the observer, and the relative humidity. Then go downstairs to the darkroom, holding the exposed film holder tight against your chest." She took me down to the darkroom. "Changing film is quite easy," she said, switching off the lights. "Turn over the film holder and shake it," she said. "The film drops into your left hand. Do you feel it? Hold it by the edge. A fingerprint could obscure a comet." She told me to run my hand along the counter until I found a drawer. The drawer held a lightproof box and a stash of film—the entire night's work. I had to put the film in the box, close the box and drawer, and load a new film into the film holder, writing the number of the exposure on the film with a pencil. All in total darkness. "Now scurry upstairs," she said. "I say scurry, because by now the observer—"

"What are you guys doing?" came faintly down the stairs, right on cue.

"Assure the observer that everything's going fine: *Everything's going fine, Gene*. It's like learning how to fly."

On my first solo I disappeared into the darkroom with a film holder clutched against my chest. I was gone for a long time, leaving Gene stranded on the telescope.

"Hurry up!" the Shoemakers called.

The Shoemakers heard a loud boom in the darkroom. Their night assistant had walked into a wall.

"Everything's going fine!" I shouted. I had accidentally switched on some blue lights, which I was afraid would expose the film, until I realized that the blue lights were inside my head. I staggered upstairs with a fresh film holder.

"Try to hold it against your chest," Carolyn reminded me.

"Okay," I said.

"Did you leave the lights out?"

"Yes."

"Did you write down the number on the film?"

"What number?"

"The number of the exposure."

"Which exposure?"

"The one you're holding."

"Uh-oh." I ran downstairs to write the number on the film.

Bangs in the darkroom told the Shoemakers that I was having further navigational problems.

When rotating the dome, the night assistant failed to note the difference between left and right. He could not locate the stop button, causing the dome to spin out of control. Night assistant read aloud a wrong coordinate, causing observer to try to search San Diego for minor planets. Night assistant violently whanged a table, belonging to the California Institute of Technology, with his head. Sounding a note of unwarranted optimism, Gene's voice came from under the telescope: "Things are shaking down after a bumpy start."

By the end of the run, the Shoemakers had exposed a total of 144 pairs of photographs. Gene developed them and hung them on lines to dry. "I just know there are Trojans in these fields," he said, "but Carolyn's got to find them." He and Carolyn put the films into glassine bags. The bags went into a box, and the box went into the trunk of the Fury. That evening the Shoemakers and their night assistant drank a small glass of white wine as a toast to the success of planet hunting, and the Shoemakers pointed the Fury at Flagstaff by the light of a waxing moon that told the end of October dark time.

Gadgeteers

August had come to Palomar Mountain, bringing a warm wind blowing through ferns and carrasco oaks and around the catwalk of the Hale Telescope's dome, where a lean figure watched the sun depart. Maarten Schmidt and his quasar team had returned for a run of four nights—their second probe for quasars. In the meanwhile, Don Schneider had been working on a computer program to scan the computer tapes for quasars, but he had not scanned any tapes yet.

Inside the data room, James Gunn, Barbara Zimmerman, and Don Schneider clustered around a computer screen, trying to get the camera on the Hale Telescope to talk to them. The computer screen came to life. It said: UTILITIES . . . ROBOTICS . . . WELCOME TO 4-SHOOTER.

Zimmerman said to Gunn, "Rewind and erase, Jim."

Gunn typed to 4-shooter: REWIND.

4-shooter said, OK.

But 4-shooter was not okay. "The tape goes out to lunch!" Gunn said. He sighed and pulled off his glasses. "What happens to it?"

"How the hell do I know?" said Barbara Zimmerman. "Try *another* rewind. Be sure and check the unit this time."

REWIND, Gunn typed again.

OK, said 4-shooter.

UNIT? asked Gunn.

10 OK, said 4-shooter.

The door of the data room swung inward, and Juan Carrasco entered, holding the box of marinated jalapeños.

"Hi, Juanito," Gunn said.

"Hello, hello," Juan said. "Where is Maarten Schmidt?"

"Who?" Don wondered.

Juan smiled. "*Doctor* Maarten Schmidt."

"Never heard of him," Don said.

"The tall gentleman."

"Oh, the tall gentleman. He's out checking the weather."

Juan put his box on a shelf, took off his hard hat, and sat at his console. Hitting switches, he started the oil pumps. He powered up a set of controls known as the phantom and the windscreen. He hit a toggle switch. The lights in the room flickered and steadied as a generator kicked in. He tapped a computer keyboard. Seven stories of steel and fourteen tons of flame Pyrex glass—the Hale Telescope—began to move slowly, visible through a window in the data room. "Just checking the weights, Jim," he said.

Maarten Schmidt entered. "Hallo, Juan. How are you?"

"Fine, Maarten, and yourself?"

"Fine," Schmidt said. "And the weather is looking agreeable." Schmidt pulled a circular slide rule from his briefcase. The slide rule was about the same age as Don Schneider. Schmidt called it—in reference to the Hewlett-Packard series of pocket calculators—"my H-P Zero." Removing his glasses and squinting at the H-P Zero, he began calculating some coordinates for the strip of sky that the team would scan tonight, looking for quasars.

The conversation turned to music.

"Somebody played the Dead Kennedys up here a while ago."

"You won't hear any of that stuff when Gunn is around."

"I bet Gunn has listened to the whole Ring Cycle in prime focus."

"Eh, probably not all of it," Gunn said. "Italian opera is very much more preferable. Now the Verdi *Requiem* is the most incredible piece of music ever committed to—"

"Jim's interest in opera is his only character defect."

OK, said 4-shooter.

Don Schneider turned to Juan Carrasco. "We want to open the dome. Is that all right?"

"You may open," Juan said.

Don went out onto the dome floor. The data room rumbled. Maarten Schmidt watched the window of the data room as the doors of the dome slit drew apart and an amethyst twilight washed over the Hale Telescope. The telephone rang, and Maarten picked it up. "Two-hundred-inch," he said, and began a quiet conversation with a colleague.

Don returned and said to the night assistant, "Juan, have you checked the dials?"

"It seems I should put in a correction," Juan said.

The astronomers were calibrating the telescope and the sensors. Jim Gunn said, "Let's go to another star."

"Moving," said Juan, hitting toggles, and a high-pitched whine carried through the data room. Dials spun. He said, "We are there." A bright star came up on the screen.

Gunn's little blue box, the kludge, still hung from 4-shooter on a piece of tape. Richard Lucinio, the digital wizard, now long recovered from his bad stomach, was rebuilding the computer that controlled 4-shooter, but the little kludge would stay taped to 4-shooter for one more run.

"How ready are we?" Maarten asked, looking around.

"Very ready," Don said.

"All right," Maarten said. "Juan!"

Juan hit three toggle switches with three fingers on his left hand, while simultaneously typing instructions to a computer with his right hand. The Big Eye gave off an *oooeee* as it slewed to the part of the sky where the quasar scan would begin. "We are there," Juan said.

Maarten said, "We have to stop the pumps, gentlemen." To Juan he added, "Gentlemen, you may stop your pumps."

"Pumps off," Juan said. He worked his way down a row of toggle switches. A whine of Vickers pumps died away, and the telescope settled and locked onto its bearings. "Phantom and windscreen off," he said.

Gunn hit a key to start the transit, and stripes flashed across the video screen. "Uh-oh," Gunn said.

Maarten peered at the stripes on the screen. "Confound it, James! This is exceedingly strange."

"4-shooter's not happy," said Barbara Zimmerman. "It got hassled."

Then the screen filled with galaxies.

The astronomers relaxed.

The galaxies were moving upward across the screen. If 4-shooter behaved, then the galaxies would move all night, while the computer recorded a strip of sky that curved in a C-shaped arc through many constellations—a cyclorama of star fields following a line of latitude around the northern pole star. 4-shooter was supposed to scan the strip automatically, without further help from the astronomers. The strip was narrow: a poppy seed held at arm's length and passed across the sky would trace a strip of similar size.

"Well, that's it," Gunn said. "Nothing more to do."

Don Schneider regarded Gunn with a raised eyebrow, and Gunn regarded Schneider with a grin. Later Don turned to the night assistant. He spoke quietly. "So how's life been treating you?"

"Very well, Don, and you?" Juan replied.

"Fine. How are Lily and the girls?"

"Fine, thank you, Don."

"Are you going to take a vacation this summer?"

"We are going to stay with Lily's mother. She has been very sick."

"I'm sorry to hear that," Don said.

Maarten took his glasses off and leaned forward with his nose nearly against the screen, until the light of the video gave his face a bluish tint. "This is fantastic!" he said. "*Things* are going by!"

"Pretty good," Don said.

"Pretty damned good!" The Principal Investigator was in fine spirits tonight. If 4-shooter kept going, Schmidt might get some quasars.

Don remarked, "We've never been able to see one quasar go by." Quasars were rare, and they looked too much like stars; it would take a computer to find them.

"It would be agreeable to see a quasar," Maarten said. He was so tall that he could not get comfortable in a chair, and so he stood up and began to walk around the data room. He spun on his heel and said, "This is a crazy kind of observing."

"Not very astronomer-intensive, is it?" remarked Jim Gunn.

"The normal standard procedure around here," Maarten said, "is running and screaming. Tonight you could hear a pin drop. If I had a pin, why, I would drop it." He put a finger on a cloud of galaxies passing by. "These things are going lickerty-snick-bit . . .

lickerty-snip-bop." He turned to Don Schneider: "What's it you say?"

"Lickety-split," Don said.

Not long afterward the screen began to blink like an advertising sign, and the astronomers began shouting in frustration.

"Confound it!" cried Maarten.

"What is this?"

"This looks like Las Vegas."

Gunn held his hands over the keyboard, fighting an urge to tell 4-shooter to quit fooling around. He had learned, however, that every time he interrupted 4-shooter's scanning with a command, 4-shooter would go dead.

Don said, "We'll have to get Jim a dummy keyboard, to give him something to play with."

"That won't work," said Barbara Zimmerman. "He can always find a way to lunch the system."

A rich cluster of galaxies erupted from the bottom of the screen— a pack of elliptical galaxies resembling a scuba diver's bubbles.

"Wow," Maarten said dreamily. "All these phenomena. All that stuff floating around out there."

A barred spiral drifted by. A white needle passed—a spiral galaxy viewed edge-on. A nearby star painted a white cross on the screen— the star had blasted the CCD sensor chip, splashing white rays across the screen. A sparkle of light glimmered. Don Schneider pointed to the sparkle and said that a cosmic ray had hit the sensor chip there. "It was probably a muon," he said. Faint field galaxies passed, cosmic confetti, wind-driven snow.

Galaxies may or may not be the basic organizational structure in the universe—no one really knows. Perhaps a majority of astronomers today believe that galaxies are *not* the major units of mass in the universe. The dark matter, or missing mass, discovered by Fritz Zwicky might be almost anything. For all anybody knows, the dark matter might include swarms of Jupiter-sized planets. The Milky Way might be stuffed with planets. The dark matter might be comets, or it might be shards of iron. It might be linear defects in spacetime known as cosmic strings. Jim Gunn often referred to the dark matter as "the stuff." That term suggested how much Jim Gunn knew about the dark matter. One fashionable candidate for the dark matter is the Wimp. The Wimp is hypothetical. The Wimp might be some kind of Weakly Interacting Massive Particle. Wimps might be able to flow through matter without touching it. Wimps might collect in invisible clouds around galaxies. On the other hand, the stuff, whether it be Wimps or not, might not be connected to galaxies at all. If the immense voids among superclusters contain the dark matter, then no one can say what the dark matter is.

Galaxies dapple the sky more richly than foreground stars in the Milky Way. While all but a few galaxies (Andromeda, the Clouds of Magellan) are invisible to the naked eye, galaxies are apparent to telescopes. The sky is a pointillism of galaxies. The disk of the full moon covers a minimum of twelve thousand galaxies. Among any given field of twelve thousand galaxies there might be—at most—about twenty-five quasars, many of which would be too faint to be detected during a scan. Maarten had estimated that several nights' worth of scans might possibly yield two deeply redshifted quasars—the remotest ones—or none at all. Scanning for quasars was like rolling dice: the laws of chance came into play, and you never knew what might pop up.

"Well, I'm off," Barbara Zimmerman said. Once again she had helped Gunn to establish contact with the robots inside his camera. Now Gunn could talk to the robots by himself.

"So long, B. Z.," Gunn said.

"Drive carefully," Schmidt said to her. "It's getting late."

There was a lull while the astronomers stared at the screen in silence. Juan Carrasco pulled one of several notebooks from his box of marinated jalapeños and made some notes in it. He felt that the only way to begin to guess what was going on inside the Big Eye was to keep track of its vital signs. He felt that the Big Eye had its good nights and its bad nights. On the first day that he had reported to work on Palomar Mountain, he had written on the cover of an empty green notebook: "Love and Ambition are the wings to success. 1969."

He had been afraid that he would fail—that he would crash the telescope. His old fear still touched him once in a while. He tried not to think too hard about the glass giant, moving out there in the darkness. The green notebook showed signs of much use. He had had to repair it with packing tape, Palomar Glue.

Other notebooks had followed the green notebook. While at first he had stuck to critical information ("astronomers' favorite radio station: KFAC 92.3 on the dial"), he had also wondered: "What happened at the moment of creation? How did the stars and galaxies come into being? How will the universe end?"—jotting questions for Jim Gunn, hoping that Gunn could answer them. Gunn, however, had been working fiendishly for most of his life to answer these very same questions, without ever attaining satisfactory answers, because (Juan noted) "What we have here is a fundamental problem."

On a shelf within easy reach, Juan placed a tattered dictionary, and when he heard a savory word, he looked it up to get the nuances. Some of the astronomers seemed to forget that the night assistant was taking notes. When they spoke of their fellow astronomers, he recorded what he had heard:

Goon—a man hired to terrorize or intimidate opponents
Yokel—a rude, naive, or gullible inhabitant of a rural area or
*   small town*
Jargon—unintelligible language or words
Grandiose Conclusion

When the astronomers saw something spectacular on the video screens, he made a note of it for posterity: "Supernova!!!"

He also kept a set of official chronicles in a gigantic red-and-black book of medieval appearance, known as the Observatory Log: "Scattered cirrus, moderate NW wind. Dr. Richard Preston (journalist) 30 yrs. old."

The box of marinated jalapeños accumulated emergency gear: seven Duracells and two extra flashlight bulbs. Two rolls of tape and some string, which he would lend to astronomers. One fever thermometer (the astronomers were careless about their health, and it was sometimes necessary to take their temperature). One bottle of Campho-Phenique ("Very good for fever blisters," he said). The marinated jalapeños box also held numerous Polaroid snapshots of beautiful objects: a ring nebula in Cygnus; a pair of nameless interacting galaxies; a comet named 1983d, which will not be seen again by human eyes until the early summer of A.D.

3027. The box held a religious booklet, bearing a message from John Greenleaf Whittier to humanity: "Nothing before, nothing behind: The steps of Faith fall on the seeming void and find the rock beneath."

Juan went downstairs to gather a midnight snack in the dome's kitchen. He returned with a tray of steaming coffee mugs and cans of soda pop and plates holding toasted English muffin sandwiches packed with a yellow mucoid that the astronomers referred to as plastic cheese.

"*Doctor* Schmidt," Juan said.

"Thank you, Juan." Maarten took an English muffin and a mug of coffee and stood up. He said, "Some part of my record player at home needs to be repaired. I am looking forward tonight to some loud, flawless music," and headed for the stereo. Fragments of rock music drifted through the data room as he scanned the dial, until *The Goldberg Variations* came up softly.

"Doctor Schneider," Juan said.

Don took a can of Von's Lemon-Lime soda and an English muffin with cheese. He did not touch coffee or alcohol, but his consumption of plastic cheese positively alarmed the cooks at the Monastery.

"Professor James E. Gunn," Juan said.

"Thank you, Juanito." Gunn accepted a can of Von's and spritzed it without removing his eyes from the endless saraband of the galaxies. He took a long swig while groping through a pile of papers until he had located his personal jumbo value-pack of M&Ms, out of which he pulled a handful of chasers to the soda.

Juan sat down at his own television screen with a cup of coffee. He sipped it thoughtfully, looking into the universe.

Don leaned back in his chair with a huge grin on his face. "Well, what do you think of this, Juan? Is this any way to do astronomy?"

Juan took a moment to consider the question while he sipped his coffee. "Yes," he said.

"All of us standing and gaping," said Maarten. He pumped up the volume on the stereo, and *The Goldberg Variations* filled the data room, Maarten Schmidt conducting with a coffee mug. "Fantastic," Maarten said. "Fantastic! It's a Big Eye, by golly! Who cares about our own eyes when we've got a Big Eye!"

▪ ▪ ▪

As an object in its own right, the universe resembles a sponge of rising dough, in which superclusters of galaxies interfinger around voids or cavities, as if the superclusters were the matrix of the sponge. As an object, the universe also looks remarkably like a swollen, pocked, filamentary cloud dissipating after an explosion— like something that began with a bang. Unlike a classical explosion, the Big Bang had no expansion center or point of origin. The explosion did not begin in any particular place. It happened everywhere out there. The prevailing theory of the Big Bang is called the inflation theory. According to this theory, at the moment of the Big Bang, the observable universe—all of the matter that makes all of the galaxies—occupied a volume of space smaller than a quark, which is the smallest known subatomic particle. The matter in the Milky Way occupied that space, along with the matter that makes up the most distant observable superclusters and quasars. During the Big Bang this microscopic, tightly compressed region of space smaller than a quark suddenly inflated into an unimaginably hot object the size of an apple, which has been expanding at a more leisurely rate ever since, until it has evolved into the present-day universe—a cold vacuum speckled with glowing floccules of matter. The universe may continue to expand or it may not. The galaxies may leave each other or they may not.

The explosion that created the universe happened somewhere between ten and twenty billion years ago. A radio telescope can hear a faint whisper of the creation. Radio telescopes can collect a signal from an event that occurred about 250,000 years after the Big Bang, when the universe consisted of a dense, hot gas. As the gas expanded and cooled, the entire universe released a sea of orange light. The light has not disappeared; it is still arriving at the earth, streaming out of the early universe from every direction in the sky. This orange light is now so deeply redshifted—coming from so far away—that it appears as the microwave background radiation, a surface of microwave emission visible all over the sky. It really is a surface: astronomers call it the surface of last scattering. The creation is visible out there. The distance from the Milky Way to the absolute horizon of our knowable universe is somewhere between ten and twenty billion light-years, although nobody is sure of the precise distance. But somewhere out there lies an image of the beginning of time, beyond which nothing, in principle, can be seen. A telescope cannot look past the beginning.

Jim Gunn took me up to the top of the Hale Telescope's dome one night, to inspect the universe. We climbed a ladder that curved up along the inner edge of the dome slit. Gunn pushed up a hatch and pulled himself through the hatchway. I followed; and we were standing on a small platform at the vertex of the Hale dome. The platform was plastered with ice, and a cold wind was blowing. "Don't fall," Gunn said, "it's a long way down." The view—pine-blanketed ridges, a yellow glow coming from Los Angeles over the horizon, the Big Dipper hanging upside down in the north— provoked thought. The earth seemed a solid place, not an oospore lost in the oceans of galaxies. I asked Gunn, "If the Milky Way were the size of a dime, how big would the universe be?"

"You mean, out to the horizon?" he asked. He was silent while he evidently calculated some numbers in his head. He said, "Incred-ible. On that scale the horizon of the universe would only be about four miles away. That's not very far at all." The wind picked up. His hair whipped around. He said, "It shows that the observable universe is a surprisingly small object. This universe—or at least the universe we can see—is a small object, something like a cloud of dimes four miles in radius." He gripped a railing and raised his voice. He said, "Essentially we live in a small watering hole."

▪ ▪ ▪

The object of the quasar search was to learn something about the edge of the watering hole, and the tool to accomplish that was Gunn's favorite gadget. Unfortunately, 4-shooter's number-two camera broke down one night. Gunn happened to be in New Jersey at the time, and so he took a taxi to Newark Airport before sunrise. Don Schneider and I both went with Gunn in the taxi, and all I can remember of the ride is the two astronomers talking about cosmic strings and gravitational lenses while a brown sun rose over the oil tanks of Elizabeth, beside the New Jersey Turnpike. I didn't take the same flight as Gunn and Schneider. By midafternoon I had arrived on Palomar Mountain in a rented car, where I found

Gunn already at work in a "garage" built into the wall of the dome beside the telescope, where 4-shooter sat prepped for surgery. It stood upright on a hydraulic cart, a white cylinder with a black lid. A scaffold surrounded it. Gunn paced underneath the scaffold, collecting tools. From his shirt pocket he removed a vinyl pouch stuffed with pens. He said, "It's exceedingly dangerous to lean over 4-shooter with one of these pen things in your pocket. It could fall down in there." He took surgical inventory. Crescent wrench. Multitest meter. Swiss officer's knife. Eyeglasses—extra-magnifying, Woolworth-type. Bag of Allen wrenches. He tucked a roll of wiring diagrams under his arm and hauled a portable oscilloscope up a ladder onto the scaffold. He dropped the diagrams on the floor of the scaffold and never looked at them again.

Gunn inserted an Allen wrench into the camera's black lid and twirled the wrench with his index finger. One by one he removed a set of Allen bolts around the top of the instrument. He said, "I need your help. We want to lift off this cover. Grab here. Pull straight up." We lifted away a hood. "Careful," he said, "we don't want to hit the cameras." We put the hood on the floor of the scaffold. Flaps and curls of Ensolite foam poked out from the interior of 4-shooter. Pipes snaked hither and thither, wrapped with foam. He said, "Are you familiar with Ensolite? It's the stuff backpackers use for mattresses. It tends to crack at minus 193 Centigrade, but it does have the tremendous virtue of being *cheap*." Some of the Ensolite-wrapped pipes delivered liquid nitrogen to the cameras, and other pipes sucked air out of the circuitry—4-shooter's light-sensor chips could operate only in the cold and vacuum of deep space.

He undid some screws and lifted a nitrogen tank away from the back of the broken number-two camera. Now we were looking inside the camera. It was a Schmidt telescope the size of a coffee can and packed with circuitry, a generous portion of which was of Gunnish design. He pointed to a small gold-plated can in the center of the camera and said, "The chip is sealed in there."

The gold can contained a silicon chip known as a charge-coupled device—a CCD. An astronomical CCD is about one hundred times more sensitive to light than the most sensitive photographic film. Many CCDs are classified secret by the United States Department of Defense, because they are used in spy satellites. The Keyhole-11 spy satellite uses a CCD; it apparently can resolve objects two to four inches across from more than one hundred miles away. CCDs will, presumably, be the primary sensors at the heart of any system of orbiting weapons. At Los Alamos National Laboratory, scientists were until recently building CCD cameras that had a working life span of a millisecond—the camera was vaporized while recording the infant face of a nuclear fireball. The CCDs inside 4-shooter can see light in the optical band; they are sensitive to the same wavelengths as the human eye, except that a CCD sees in black-and-white rather than in color. The human eye can discriminate among sixteen levels of gray. A CCD can discriminate among six thousand levels of gray. 4-shooter's CCDs were made by the Texas Instruments Corporation, and although they are among the most sensitive CCDs in the world, they are not classified. Texas Instruments built them under a contract with the Jet Propulsion Laboratory, in Pasadena, to be the sensors inside the main imaging camera on the Hubble Space Telescope. This camera is called the Wide Field/Planetary Camera. It so happens that Jim Gunn is one of the principal designers of the Wide Field/Planetary Camera, which is why he was able to get his hands on four supersensitive CCD chips. Each chip is square, and 1.2 centimeters on one side. The chip is a pure, flawless, translucent crystal of silicon, and is extremely thin. Fifteen of these chips, if pressed together, would add up to the thickness of a piece of paper. If you blew a breath of air onto a CCD chip, it would shatter. The chip is covered with a grid of sensors that are known as pixels. The grid on the surface of a Texas Instruments CCD contains 640,000 pixels and no less than forty feet of microscopic wiring.

A CCD is difficult to manufacture. The sheet of crystalline silicon must be etched with acids to a thinness greatly exceeding that of sandwich wrap, and the crystal has to be perfect, because a single defect in the crystal will render the chip useless. The Texas Instruments CCD group, under the direction of a solid-state physicist named Morley Blouke, fabricated about twenty-five thousand of these chips in order to get about 125 chips that worked reasonably well, of which eight were selected for use in the Space Telescope's Wide Field/Planetary Camera. Gunn's Texas Instruments chips might have cost fifty thousand dollars each if Gunn had paid for them, but he managed to scrounge his four chips for nothing. They had slight flaws. They were, in a word, surplus parts. The CCDs inside 4-shooter would never work inside the Space Telescope, but chilled with liquid nitrogen and encouraged by gobs of circuitry, they work nicely enough on earth. Coupled to the Hale Telescope, 4-shooter could see a lit cigarette seven hundred miles away.

Gunn put on his Woolworth glasses. He found a mechanical pencil. He pointed the pencil at a wire emerging from the gold can that held the CCD. "Can you see this wire?" he asked. The wire was as thin as a human hair. "That's the video wire," he said. "It brings the signal out of the chip and into the circuitry. The chip talks to the outside world through that wire." He remarked that the video wire was brittle. "If you touch the video wire, you break it."

The CCD was hermetically sealed inside the gold can. The can had a window through which the chip looked out at the universe. The liquid nitrogen tank on the camera had a cold finger that touched the can and cooled the chip down to nearly the temperature of liquid nitrogen, which made the chip exquisitely sensitive to light. A CCD chip collects light just as photographic film collects light. When 4-shooter's four camera shutters are opened, starlight falls onto the four chips. The chips can be left exposed to the sky for several hours. During long, deep time exposures, light coming from a galaxy may arrive on the surface of a chip a single photon at a time, at a rate of one photon every few seconds. When photons impinge on the silicon crystal in a CCD, they knock electrons out of silicon atoms. The electrons collect in the grid of pixels on the surface of the chip. After an exposure, the camera shutters are closed. 4-shooter's computer system then pumps the electrons from the pixels, causing the electrons to flow out of the chip, over the hairlike video wire, and into a set of amplifiers inside 4-shooter, where the electrons are counted and converted into digital numbers. From there, the numbers flow through cables into video monitors in the data room, which fill with wraiths of nameless galaxies, and finally onto computer tapes. Ever since Gunn had built the little blue kludge box, 4-shooter had been able to pump out unbroken rivers of electrons while the sky moved past the mouth of the Hale Telescope and microscopic images of galaxies crept across the faces of the chips.

Gunn rolled up his sleeve and slipped a metal band around his arm, to drain off any static electricity in his body. If his body were to contain any static electricity when he touched the circuitry, stray electrons might travel backward up the hairlike video wire, into the CCD chip, and fry it. If you ran a comb once through your hair and then touched the comb to a CCD, that would vaporize wiring all over it, destroying it. There had been plenty of accidents with CCDs. Before they learned the delicacies of CCDs, gadgeteers had sneezed over them, shattering the chip or spraying it with saliva—in either case, a fifty-thousand-dollar sneeze.

Gunn stared into the circuit boards of the number-two camera. He said, "Whatever the problem is, it is in here," and started poking at things with the eraser on his pencil. He clamped a probe into a circuit board and studied a jagged line on the oscilloscope. He wiggled a contact with the eraser. He bumped a transistor. He watched the line on the oscilloscope. He muttered. He craned his neck and peered into the camera.

Suddenly: "Jesus H. Christ! It's a loose washer."

A washer had dropped into a circuit board and shorted it out. Fingering a pair of alligator pliers, he delicately lifted the washer free. "Would you look at that," he said. Then he stared off toward the Hale Telescope, holding the washer in the pliers, trying to imagine where that washer had come from. He said, "That washer was sitting *exactly* where it would do *exactly* what it did." Which, he added, was to cause the video screens to go blank.

Where there is a loose washer, there has to be a loose nut. In a somewhat dejected tone of voice he said, "There would be three other washers loose in here too." He peered into the camera. "Okay. I see the nut. I see *two* washers. Now where is that last washer?" He poked around with the pliers. "How the hell do I get them out of there? There must be some long, skinny tweezers around here somewhere." He scrambled down the ladder. He returned with long, skinny tweezers and a flashlight.

While I held the flashlight, he slid the tweezers into the camera. "Here's the nut," he said, holding it up. He plucked the two washers out of the camera. Then he spotted the third and last washer deep inside the circuitry. "What we need," he said, "is a sticky probe." He ran down the ladder and returned with a roll of Scotch tape and a length of stiff wire. He wadded a ball of Scotch tape around the end of the wire. That was the sticky probe. Then he pushed the wire down into the camera, trying to snag the washer on the sticky tape ball. *"Dammit,"* he said. The tape ball had fallen off the wire. Now he had a CCD camera with a washer *and* a ball of Scotch tape lost inside it. "The sticky probe wasn't going to work anyway," he said. "Unfortunately I am going to have to take this board out of here. If you could turn the flashlight this way . . ." Working deftly with the pliers, he pulled out a circuit board and retrieved the last washer and the ball of Scotch tape. Then: "Damn. I broke the video wire. That was inevitable."

Now he had to solder a new video wire. He went downstairs to the electronics shop, where he lit a propane gas heater to warm up the shop, and put on a different pair of Woolworth's reading spectacles, in order to see the video wire. He held the video wire close to his face and dabbed at the wire with a soldering iron. "You want an instrument so badly," he remarked, "that finally you have to go and build it yourself." He turned the wire in his fingers and poked it into a droplet of molten solder.

Gunn's hands displayed cuneiform marks left by molten solder that continually rained on them. "Fortunately solder burns aren't permanent," he said. His left index finger was slightly crooked and stiff, prone to mild arthritis. His knuckles were large and knobby, but the tendons that controlled everything were as taut as piano wires. His left thumb bore a scar—"the result of an accident, which I do *not* like to think about, with a wire brush on a power grinder." The wire brush had shredded the inside part of his thumb down to the bone.

"I cannot *bear* to see something not working," Gunn once said. During cloudy nights on Palomar he had taken apart and fixed cars and trucks. He had torn down and rebuilt a gearbox that drove the dome on the Oscar Mayer Telescope. He had repaired the old Otis elevator in the Hale dome. The department of astronomy at Princeton owned a powerful computer that crashed one day, and not even the manufacturer's service representatives could figure out what was wrong with it. Gunn started poking around. He pulled out a circuit board packed with chips and soldered in its place a single resistor that had cost around fifty cents. The computer came back to life, apparently no worse for its chipectomy.

Gunn had built himself a stereo—an awesome kludge. He had begun it as a hi-fi in 1964, and after decades of mutations Gunn's stereo had almost completely renewed its material substance, just as the human body is said to replace itself every seven years, although Gunn's stereo still contained a few 1964 amplifier tubes, just as the human body contains teeth. The stereo had fattened into several cubic feet of components, until it didn't fit well in Gunn's living room. Gunn had finally had to bury it in a crawl space under the living room floor in order to get it out of the way. During the times when he was off traveling and Jill Knapp, his wife, wanted to hear music, she had to spend ten minutes toggling data switches and plugging and unplugging coaxial cables that led under the floorboards, trying to get sound out of Jim's stereo, because its wiring diagram mysteriously seemed to change every time he played music on it.

The fact that he had to build his own scientific instruments frustrated Jim Gunn, because he was a theoretical cosmologist as well as a tinker. He had never been able to settle on a career. He did not know whether he preferred to ponder the nature of the dark matter (quark nuggets? fast-moving cosmic strings? rotted Wimps?) or to fiddle with sticky probes. So he did both. He had written 200 papers, mostly with coauthors, on subjects ranging from the evolution of galaxies to the design of new machines. "Most of the science I'm doing now is highly collaborative," he said. "Inevitably I play the role of engineer, because inevitably it is my instrument we use, which inevitably needs care and feeding."

The world of astronomy held three types of people: observers comfortable with telescopes; theorists comfortable with pencil and paper; and instrument builders comfortable with wires. Mother Nature valued her secrets too highly to let any astronomer run wild in all three fields, but Mother Nature had somehow overlooked Jim Gunn. A writer named Jim Merritt once called Gunn "a kind of triple threat ranging across all three fields," a line that reverberated among Gunn's friends, who nicknamed him the Triple Threat.

Some years ago the following sign appeared on a bulletin board at Caltech:

GUNN FOR A DAY CONTEST!
In 500 words or less, write an essay describing why YOU
want to be *Jim Gunn* for a day.

GRAND PRIZE
For one day, you can be Jim Gunn!!!
You will be magically enabled to:
1) Write one paper that is a major theoretical breakthrough.
2) Design one instrument.
3) Successfully avoid your graduate students all day.

Being Jim Gunn for a lifetime brought much excitement but very little peace. "I don't end up with much time to think about what it all means," he said.

The boss of the Wide Field/Planetary Camera project, Jim Westphal, who has worked with Gunn for a decade, once said of Gunn, "Not only does this man do wondrous great science, he helps others do wondrous great science. He is such a kind man that he bails everybody out of their problems. As long as he's bailing me out, I think that's wonderful." Gunn had memorized the cabala of exotic parts. Jill Knapp, his wife, once spent a day designing a unit for a radio telescope, and then showed her design to Jim. "You can buy that for sixty-nine cents," he said, and gave her an address.

He had a tendency to drop out of sight once in a while, drawn, perhaps, by the aroma of burning solder drifting from a basement laboratory somewhere. Jill could not always track Jim once he had gone to ground. "I have lost Jim for twenty-four hours at a time," she said. He spent so much of his life on airplanes that one of his graduate students said of him, "Jim Gunn could be defined as a probability function that peaks over the middle of the United States." There were widespread doubts that Gunn ever slept. "Yeah, he sleeps," Jim Westphal said, "but he doesn't sleep very often." Concerning his own situation, Gunn once remarked, "I feel incredibly privileged to be able to do something in life that is this much fun." One of Gunn's fellow Princeton astronomers, Edwin L. Turner, thought, "In some imaginary sense, astronomy might be better off if Jim Gunn were three people." So might Jim Gunn, but since the laws of physics limited him to one point in space and time, he had to travel endlessly around North America, somewhat dismayed by the fact that in the ages since the Big Bang, the average spiral galaxy had rotated on its axis at least forty times—had existed for at least forty galactic years—evolving, as it turned around and around, into a hypnotic, heart-pounding wonder, whereas he, as a temporary collection of proteins, would remain intact for what amounted to only ten seconds out of a galactic year—not exactly enough time to figure out what it all means.

James Edward Gunn, Senior, and his wife, Rhea, had their only child in 1938, in Livingstone, Texas. They named him James Edward Gunn, Junior, and one of the things he had in common with his father was the eyes: brown and mobile. When the boy grew older, those eyes could be by turns piercing or shy. James Gunn, Senior, had a square jaw, a bold forehead with heavy eyebrows, and brown hair (what was left of it). Jim Senior looked a lot like his son does today, except that he was clean-shaven and wore a suit and a felt hat. Jim Senior was very nearsighted and wore rimless spectacles. He loved science, and if it hadn't been for the Depression, he might have become a college professor; but instead he became a wandering oil prospector. He worked for Gulf Oil Corporation, looking for pools of crude oil with gravity meters. Every year or so, Jim Senior moved with his crew to a new site, taking Rhea and Jimmy with him.

Jim Gunn once narrated his childhood to me, in these words: "First half of first grade, Chipley, Florida. Last half, Meridian, Mississippi. The whole of second grade, Bossier City, Louisiana. Third grade and half of fourth, Plainview, Texas. Hum. What. Oh, Christ, I thought I could reel these towns off. The other half of fourth grade seems to have vanished—I don't know where we were living then. Anyway, when I was in fifth grade, we moved to Camden, Arkansas. Life was a little strange. One of the bad things was that we never stayed anywhere long enough for me to make close friends. I'm rather grateful in a way—most children are affected by cliques and I was not."

When they moved, Jim Senior packed his tools into a clamshell trailer and towed it behind the family Ford. "During World War II and just after, running an oil exploration crew was difficult," Gunn recalled to me. "You had trucks and machines, but you couldn't get parts. So my dad manufactured his own parts." Jim Senior had made the trailer too. He shaped it from aircraft aluminum into an ellipsoid with flat sides. The trailer opened along hidden lines—its front and stern cracked apart and lifted up like wings. Shelves dropped down and secret doors unvalved, revealing a drill press, workbench, power hacksaw, band saw, grinder, plane, vise, lathe, electric welder, acetylene torch, and racks of hand tools—a Texas oilman's Fabergé egg.

"I got most of my early education from my dad," Gunn explained, and it was not what was taught in the public schools. "I got my hands dirty with lathes." Jim Senior rolled up his sleeves when he worked in the aluminum egg alongside his boy, and he taught his son a love of building things with his hands. "My dad," Gunn said, "was really my buddy."

Jim Senior liked to say, "My main excuse for having a boy child is so I can have an electric train." The two of them created a piece of Virginia, sculpting Appalachian ridges from papier mâché, shot through with tunnels. They built a town and a switchyard, and assembled HO gauge railroad cars from zinc parts. The boy named it the Shenandoah Valley Lines. When Jim Senior moved the family, he and his boy pulled up the towns and piled the mountains into boxes. Like the aluminum egg, the Shenandoah Valley folded for travel.

When Jim was seven years old, his father gave him a book on astronomy, *The Stars for Sam*. He read it so fast that Jim Senior gave him another book on astronomy, and he read that one, too, from cover to cover, when he was seven years old. It was a college textbook. After that, with the help of the aluminum egg, Jim Senior helped his son build a small refracting telescope. They made a body from a mailing tube. They took a piece of glass from a pair of spectacles and ground it to fit the tube. The lens was one and a half inches across, the same size as the lens on Galileo's telescope. The view from Bossier City, Louisiana, wasn't bad. They saw craters and seas on the moon. When they pointed the mailing tube at Jupiter, they saw a salmon egg in the sky, attended by tiny moons.

They pointed it at the mists of the Milky Way, and they saw the Milky Way jump into finely divided stars. Jim Senior consulted his star chart with a flashlight: Okay, you see that bright star. Move a little to the left and down. They saw Orion's Sword, where stars are catching fire. They searched for the Ring Nebula. The boy believed that he could almost see it—a bubble of gas where, he would learn later, the outer layer of a star had lifted off into space, resulting, perhaps, in the incineration of the star's solar system.

Miniature worlds and models fascinated Jimmy Gunn. The Shenandoah Valley was a closed, symmetric universe governed by laws. He also began experimenting with flight. "I started making model airplanes practically at age zero," he recalled. He worked with kits at first. With exquisite patience he would glue an airframe out of ribs of balsa wood, stretch paper over it, and dope the paper with chemicals to shrink it over the airframe. Then he would bolt an alcohol engine to the plane's nose. To get the plane aloft, Jimmy and Jim Senior would take it to an open field and wire it to a car battery. Jim Senior would flip the propeller until it caught with a sound like an elf's chain saw, and let the plane go. The boy held a control line, turning around and around while the plane circled and Jim Senior cheered. He could make it climb or dive by working the line. Sometimes the plane would hit the ground and break up into fragments—weeks of work gone in a burst of splinters.

Jim Senior was a private man who did not share all of his life with Rhea or his son. He would put his hat on his head and drive off for a few days without telling them where he was going or why. Rhea got hints that he was visiting east Texas, maybe Houston. Jim Senior never talked about it. In the summer of 1949, Jim Senior moved the family to Camden, Arkansas, where he prospected for oil and gas with his gravity meters. Jim started fifth grade in Camden. At Christmas that year, Jim Senior loaded his boy with presents. He gave Jim an Olson 23, one of the best airplane engines that money could buy. He also gave Jim a kit for a reflecting telescope with a four-inch mirror, a beauty. They made plans to build the telescope together. One day in February, Jim came home from school to find an ambulance parked in the driveway beside the aluminum egg. The house was full of people. "An aunt of mine from Shreveport was there, and she whisked me away and told me as gently as she could." His father had died of a sudden heart attack. Jim Senior may have known or suspected he was dying, which might explain all the magnificent presents he had given his boy that Christmas. And those unexplained trips to east Texas had been to see a heart specialist. His heart had been failing on him, but he hadn't wanted to tell his wife and son. The ambulance pulled out, bearing the body of Jim Senior.

As Gunn recalls, "It hit slowly, over the next two years. My dad was a key part of my world, and that world became empty for a while after he was gone. He was not there anymore, and I missed him terribly. I still miss him. In truth I don't like to think about all this very much, because it is still a very painful thing for me."

Rhea moved Jim and herself to Beeville, Texas, a small community in the south coastal flatlands behind Corpus Christi, where her sister lived—a country full of towns with names like Pettus, Refugio, Mineral, Goliad, and Poth. Row crops did poorly in the hard soil, except for some cotton. Cattle and the military did better; Beeville had a naval air station. The town had two main streets and four or five blocks of clapboard houses, which sat on feet made of cinder blocks. Jim and his mother moved into a little white duplex, and she found a job as a clerk in a drugstore. For two years they lived in Beeville, the longest time Jim had stayed anywhere. They were not destitute, since Jim Senior had had some insurance. Rhea told Jim that he could have a room of his own built onto the back of the house. Jim laid out the specifications: many shelves for books and tools, and a workbench. But they had to sell his father's clamshell trailer workshop, the aluminum egg. "It had deteriorated somewhat," he remembered, "and I was too young to keep it up. That was one of the great sadnesses." For the rest of his life Jim Gunn would feel a sense of loss whenever he recalled his father's aluminum egg. He did not know what had happened to it; probably junked. He figured that if he only had it now, he would park it next to the Hale Telescope. He said, "I could use it right now."

There were other ways to fill an empty planet. He designed and built twenty or thirty experimental airplanes in his room at the back of the house. Some lofted into the blue, went out of control, and dived at seventy miles an hour into the ground. Rhea married an Army man, Bill Taylor, and Bill took Rhea and Jimmy to New Boston, Texas, near the Army's Red River Arsenal in Arkansas, where Bill was stationed.

At the Red River Arsenal the Army provided a woodworking shop inside a Quonset hut, "to try to keep the soldiers off the bottle," Gunn now thinks. He asked his stepfather if the two of them might go down to the Quonset hut on weekends and build a telescope. Bill Taylor agreed, because he wanted to be a good father to this boy. Jim then sent away in the mail for a cheap lens the size of a salad plate. He and Bill Taylor built a six-foot tube for the lens, in the Quonset hut, and the soldiers marveled at the two of them, constructing what was apparently the biggest telescope in Arkansas. When Jim and Bill let first light into it, they found out that it was also the worst telescope in Arkansas—the mail-order lens was no good. Then the four-inch mirror that his father had given him, that last Christmas, came into Jim's mind, but the kit did not include a tube. Jim found a rainspout and cut it to the right length and built a second telescope in the Quonset hut. He took it out at night, and he could see colored bands on Jupiter. Then he built a third telescope, a refracting telescope with a three-inch lens, and with it he could see the Ring nebula in Lyra.

The Army transferred Bill Taylor to Okinawa. Jim and Rhea moved back to the house in Beeville for the time that Bill was stationed overseas, during Jim's high-school years. Jim sold the Shenandoah Valley railroad model when they moved; it had become troublesome to pack. Today Gunn wanted those trains. He admitted that their disappearance was another of the sadnesses. Back in his room in Beeville, he built furniture. He discovered classical music simultaneously with vacuum tubes and built himself a hi-fi. He also acquired a friend by the name of Bill Davis, and the two of them hit upon a new way to reach for the stars. They sent away in the mail for zinc, sulfur, and potassium perchlorate. They mixed the chemicals and packed them into finned steel tubes, followed by a plug of gunpowder for an igniter. The launch zone was a cow pasture outside Beeville. Most of the rockets flashed and fizzled on the pad, which unnerved the cows, to be sure, but not as much as when a rocket rose a hundred feet into the air and burst in a hail of shrapnel, which terrified the cows.

Flying pipe bombs intrigued them, but liquid fuel inflamed their imaginations. "We knew what was going on at White Sands," Gunn says. At the White Sands Proving Grounds, Wernher von Braun was firing off captured V-2s and liquid-fueled Aerobees. In a sort of homage to von Braun, Jim Gunn and his friend Bill Davis sent away in the mail for bottles of liquid aniline and nitric acid. When they poured the aniline into the nitric acid, something agreeable happened: the mixture spontaneously burst into flames. But mail-order nitric acid, they sensed, was too gimpy for a rocket motor—no kick. They decided that they would need something called red-fuming nitric acid if they wanted a real high-powered launch. That stuff had kick, which was exactly the reason why you could not order it through the mail. So they built a distilling apparatus in Gunn's room, a tangle of glass bulbs and pipes, and with it they manufactured nitrogen dioxide gas, which they bubbled through a glass vessel full of nitric acid. The nitric acid turned red and belched—it became red-fuming nitric acid. "It made an angry red cloud," Gunn remembers, which was highly toxic, although not as toxic as the aniline, which was a contact nerve poison. Then they built a rocket motor. It consisted of two fuel tanks and a rocket nozzle. They mounted the motor on a bench in order to test it. They did not put any kind of an igniter on the motor, because they suspected that the motor would not need any help getting lit. Davis's father owned a sporting-goods store. They set up the motor on its bench in an open lot behind the store, for a static test-firing and thrust analysis. They pumped one tank full of pressurized aniline and the other tank full of pressurized red-fuming nitric acid. They cracked open the valves leading to the rocket nozzle, and they heard a sound like somebody tearing a bed sheet in half, and the nozzle belched a fireball that swallowed the nozzle, the tanks, the whole bench. They ran for their lives. It wasn't a rocket motor, it was an industrial accident. A mushroom cloud boiled up from behind Davis Sporting Goods.

Then he discovered astronomy. "Astronomy for me was a lone pursuit," Gunn says. "Once I got into mirrors, I stopped playing around with explosives." He wanted to see galaxies. He sent away in the mail for a blank disk of Pyrex glass, eight inches across. Using a polishing tool faced with black pitch, Gunn ground the

Pyrex into a deep hollow, putting on successively finer grades of
Carborundum grit and a lot of water. Using cerium oxide polish,
he sleeked the glass to a glittering paraboloid curve. He built a
tube and a fork for his telescope. He installed the mirror in the
tube. He mounted the telescope on an oak stump in the backyard.
Photons coming from remote corners of the galaxy, and from
beyond, fell with democratic abandon all over Beeville, Texas, available to anyone who wanted to catch them with a homemade mirror.
He saw colored stars—green, yellow, blue, orange. He saw fingers
of black dust in the Lagoon nebula. The middle "star" of Orion's
Sword was a cave of gas, and at the center of the cave burned four
sapphire stars called the Trapezium. The Trapezium stars had been
born at the same time—a brood hatched in a nest. He could see
an antique glow at the core of the Andromeda galaxy—elderly
stars—and he could see two dwarf elliptical galaxies in orbit around
Andromeda.

Gunn was disappointed with his telescope. The Whirlpool galaxy,
for example, looked like a ball of cotton. He felt that the human
eye suffered from terrible design, but that a camera could capture
clearer vistas. Time exposures on Kodak film would reveal the spiral
arms of a galaxy. Galaxies rise in the east and set in the west as
the earth turns, which meant that he would have to equip his
telescope with a drive mechanism, to track the sky, and a camera,
allowing for time exposures.

This required money. His only liquid assets were the three telescopes that he had built at the Red River Arsenal. He put a classified
ad in the Beeville paper, announcing three telescopes for sale. A
passing stranger saw the advertisement. He was a carnival man.
In the 1950s, and earlier, telescope men were figures occasionally
to be seen in small-town carnivals, selling crowds a tour through
the solar system for twenty-five cents. Gunn tried to explain to the
man that you couldn't see anything through the giant telescope,
the one with a lens as big as a salad plate. Sure, the telescope man
said, but who the hell in Waco will know the difference? The man
peeled a few greenbacks from a roll of bills, and that was the last
Jim Gunn ever saw of his first three telescopes. They hit the road
with a carnival man.

The greenbacks were seed money for a photographic system that took Gunn five years to build. In the meanwhile he went to college at Rice University, where he finished the telescopic system during his senior year. The telescope had now evolved into a stark white cylinder loaded with cameras. He had wired the telescope up to a transistorized drive box crammed with electronic gear, including military-surplus parts. The telescope had two cameras: a wide-field camera and a planetary camera. The wide-field camera took wide-angle exposures of deep sky, while the planetary camera took close-ups of planets. People in some cultures believe that the souls of the dead can enter inanimate objects, such as rocks and trees. If this is true, then the spirit of Jim Senior must have entered into this telescope's exquisite gadgetry, which was as compact and resourceful as Jim Senior's aluminum egg.

Gunn photographed the Pleiades, the Horsehead nebula, the Veil in Cygnus, the Rosette nebula, and the spiral arms of the Whirlpool galaxy. The images were crisp and dramatic—*Sky & Telescope* magazine ran two stories on Gunn and his pictures. He graduated number one in his class at Rice University, a math-physics major. During his high-school years he had begun dating a Beeville girl named Rosemary Wilson, and shortly after their graduation from college, he and Rosemary were married. They moved to California, where Jim began graduate school at Caltech in astronomy.

At Caltech he became fascinated with cosmology, the science that deals with the birth, life, and death of the universe as a whole object—starting at the Big Bang and ending with the fate of matter. Gunn explored Albert Einstein's equations of general relativity, which describe, in four dimensions, various possible presents, pasts, and futures for our universe. If Gunn had been a normal human being, four dimensions might have satisfied him, but he could not get rid of a chronic disease, the *cacoethes gadgetendi*, the itch to tinker. On his own initiative he decided that what Caltech needed was a machine to analyze star images in a glass photographic plate. The head of the astronomy program, Jesse Greenstein, gave him a midget darkroom in a basement of the Robinson building in which to build the machine. The room was so small that Gunn ended up prefabricating his machine in chunks at home and assembling them in the darkroom. (I once remarked to Jesse Greenstein that a darkroom seemed a rather narrow place for someone of Gunn's ambitions. "Yeah?" he said. "Who's crapping? If you give these guys too much room, they don't produce.") One day Gunn rolled something out of the darkroom on wheels. It was a gray metal cabinet, considerably larger than Jim Gunn, and studded with fifty-four dials. The thing is now known to some people as Gunn's First Machine. It works in the following manner: You clamp a glass photographic plate into an iron stand. A sensor touches one star in the plate. The sensor picks up an image of the star and feeds it into the machine. The machine analyzes the image and declares the exact brightness of the star. Jesse Greenstein still uses Gunn's First Machine.

During the summer of 1965, Gunn asked one of his teachers, a gadgeteer named J. Beverley Oke, if he could observe on the Hale Telescope, a rite of passage for young Caltech astronomers. Bev Oke took Gunn to the mountain. Oke worked most of the night in prime focus at the top of the telescope, using an electronic instrument to collect red light from a quasar known as 3C 273. A few minutes before dawn, Oke came down and told Gunn to go up. Gunn stepped onto an aluminum platform that looked like a diving board: the prime focus lift. He hit a button, and the lift rumbled up along the inside of the dome, rising past the shadowed girders of the tube, past the curve of the enormous horseshoe bearing, until the lift came to a halt at the lip of the Hale Telescope. There, suspended in the mouth of the telescope, was a small room: the prime focus cage. This room was not unlike a lidless tin can.

Gunn stepped off the diving board into prime focus. He crouched on a tractor seat. He leaned over Oke's instrument, which sat in the center of the room. It had an eyepiece that looked downward at the mirror. He put his eye to it and saw a set of illuminated crosshairs. The night assistant uncovered the mirror, and Gunn saw a reflection of the universe, already fading in twilight, and suddenly the room tilted sideways and the tractor seat rolled, while Gunn fumbled with the controls of a starship, trying to catch the quasar known as 3C 273, chasing it down as it set into mist over the Pacific Ocean.

▪ ▪ ▪

"Those big telescopes are a little like drugs," Maarten Schmidt once said to me. He had probably spent more time at prime focus in the Hale than anyone else on earth. Schmidt had learned astronomy at the University of Leiden, in the Netherlands, and first arrived in Pasadena in 1956, having recently married Cornelia Tom, who had been a kindergarten teacher in the Netherlands. He and Corrie spent two years in Pasadena. They returned briefly to the Netherlands, and then Jesse Greenstein offered Maarten a job at Caltech. Maarten and Corrie settled in California and raised three daughters there.

A Caltech astronomer named Rudolph Minkowski retired soon after Schmidt joined the Caltech faculty. Minkowski was a supermassive astronomer who had difficulty climbing in and out of the prime focus cage, but he nevertheless had pioneered the study of radio galaxies (galaxies that emitted hissing radio noise). When Minkowski retired, he left behind an unfinished observing program, for the sky was too big even for Rudolph Minkowski. Schmidt, virtually by default, took over Minkowski's program and found himself looking at radio galaxies.

Isaac Newton (one of the original gadgeteers; Newton invented the reflecting telescope) discovered that if he passed the light of the sun through a prism, the prism would produce a patch of color that ranged from blue to green to yellow to a red as dark as blood. A prism, Newton discovered, broke sunlight into its component colors. Newton had invented spectroscopy, or the decomposition of light, which is one of the central techniques of astronomy. By the use of a prism or a mirror ruled with fine lines, the light of any star or galaxy (since galaxies are made of stars) can be decomposed into a slash of color that goes from blue to red, as Newton did with sunlight. This streak of color is called a spectrum. For the foreseeable future, the decomposition of light is the only way we will touch the stars. To make a spectrum is to collect and analyze a star's material—photons that came from the surface of the star.

A star's spectrum is brighter in some wavelengths of color, darker in others. When light from a star is spread into a spectrum, the spectrum shows black bands—narrow, dark gaps marking wavelengths where little or no light comes from the star. These are called absorption lines. They are caused by relatively cool gases and vaporized metals, near the surface of the star, that absorb light at particular wavelengths, thereby blacking out the spectrum in those particular colors. Certain stars—Wolf-Rayet stars, dwarf emission stars—show bright bands in their spectra; distinct, brilliant colors in which large amounts of light pour from the star. These glowing bands in a spectrum are called emission lines, and they are caused by hot, luminescent gases in and around the star, excited by radiation until the gases fluoresce in distinct colors, as does, for example, the gas in a neon lamp. During the nineteenth and early twentieth centuries, astronomers perfected techniques for picking apart starlight into its component colors. They learned how to identify dark absorption lines and bright emission lines as signatures of various elements—hydrogen, carbon, oxygen, metals. They passed the light of a star through a prism onto a black-and-white photographic plate, thereby producing a black-and-white banded streak. They looked at the bands under a microscope to determine the constituents of the star.

Most light is invisible to the human eye. The total spectrum of light goes from short-wavelength gamma rays, to X rays, to ultraviolet light, to visible light, to infrared light, to microwaves, and finally to long-wavelength radio waves. These are all forms of electromagnetic radiation, and thus they are light. The colors that the human eye can see amount to a razor-thin slice of the total spectrum of light. By the 1950s, it had become clear to astronomers that objects in the sky emitted much light other than that visible to the eye. Radio detectors began to reveal spots of radio emission all over the sky. Antennae in those days were not keen enough to pin down the location of a source; most radio spots were resolved only as blobs of noise, too fuzzy to be linked to any particular stars or galaxies. Astronomers felt the frustration of an ornithologist standing in a forest and hearing birds of unknown species singing in the trees. Listening to the songs of birds, the ornithologist sweeps the trees with binoculars, trying to identify new species. Some birds display themselves, but most remain hidden in the foliage. In an effort to aid the task of identification, astronomers at Cambridge

University in England assembled several lists of radio blobs. The third of these lists, which is probably the most famous, is generally known as the third Cambridge survey of radio sources. At that time astronomers thought that most sources of radio emission in the sky would prove to be either radio galaxies or threads of excited gas left over from supernovas, but nobody could be sure, since most sources listed in the Cambridge radio surveys remained unlinked to any objects that could be seen through a telescope.

In the fall of 1960, Thomas Matthews, a radio astronomer, managed to pin down the location of one radio source, 3C 48. (3C stands for "third Cambridge" and 48 indicates that it is the forty-eighth source of noise listed in the catalog.) 3C 48 was a blue star. Allan Sandage, an optical astronomer, became interested. From prime focus in the Big Eye, Sandage photographed 3C 48 and found strange colors. Measuring the object, he found it to be a point source—an object of minuscule diameter, as seen from earth. It appeared to be some kind of a radio star, or possibly the remains of a supernova. Tom Matthews found more locations for the Cambridge radio sources. Some of them turned out to be radio galaxies, and some turned out to be blue stars. A few astronomers began referring to these objects as radio stars, but in general, the astronomers who were looking at them did not think of them as any one class of object.

Jesse Greenstein decomposed the light of several radio stars into spectra, trying to figure out what they were made of. The light mystified everyone who studied it. They found inexplicable patterns of stripes—emission lines painted on top of a spectrum that glowed brightly at all optical wavelengths of light. The emission lines were soft and wide. They signified a bizarre object: something extremely hot, under enormous pressure, containing clouds of gas moving at high speeds, and evidently made of unknown matter.

Meanwhile, as he grew into Rudolph Minkowski's job, the young Maarten Schmidt began spending long nights in the prime focus cage at the mouth of the telescope, using an instrument called the Prime Focus Spectrograph to break the light of radio galaxies into streaks on photographic plates. It had a slit that allowed the light of a single galaxy, reflected from the Hale mirror, to pass onto a reflective prism. The prism fanned the light into a rainbow, The rainbow went into a camera and bounced off a mirror, passed through a lens, and hit a glass photographic plate that was the size of a fingernail. The plate was so small and frail that you could pick it up just by touching it with a fingertip, on which the plate would stick. There were two interchangeable cameras for the Prime Focus Spectrograph. One had a lens made of sapphire, the other of diamond. Ira Bowen, who had directed the final testing and figuring of the Hale mirror, had designed these cameras. One of his designs called for nothing less than a diamond lens *half an inch* across. Bowen had no idea where he would find a diamond that big for a price he could afford, but a quiet investigation led Bowen to a diamond dealer who had been using a flat diamond as a watch fob. Bowen persuaded the dealer to part with his watch fob for very little money, since the stone was too thin to be cut. Bowen gave the fob to Don Hendrix, who went to work polishing it with powdered diamond mixed with Vaseline, and turned the fob into a lens.

Just after Christmas, 1962, Schmidt went to Palomar Mountain for a run in which he planned to take spectra of radio galaxies. On the night of December 27, he spent nine hours gathering the light of a radio galaxy. Toward dawn, with a couple of hours on his hands, he turned his attention to a radio object in Virgo, listed as object number 273 in the third Cambridge catalog of radio sources—3C 273. He had seen a photograph of it. It was not a remarkable object at all—just a faint streak or a cloudy filament that was emitting radio noise. He thought it was probably a thread of excited gas. He prepared to take a spectrum of it.

First he had to load the camera. Working by feel in total darkness, he lifted up the camera's lid. He clamped a tiny glass plate into the camera. He snapped down the lid. The night assistant slewed the telescope to Virgo and centered it on the star field that contained the shred of gas, 3C 273. Maarten put his eye to an eyepiece and looked down onto the looking glass, fifty-five feet below, where he saw mists of stars, like pollen on a fish pond. He looked around. He recognized the group of stars that held the radio streak. He could not see the radio streak with his naked eye, but there was, however, a very bright star sitting next to where the radio streak ought to be. He decided to take a spectrum of the star, just to get that out of the way. Pushing buttons on a paddle, he tweaked the Hale Telescope until the light of the star came through the slit of the spectrograph. He pulled a dark slide shutter, and the exposure began.

Next, looking through another eyepiece, he selected a guide star—a bright star located somewhere in his field of view—and placed a set of crosshairs on it. Whenever the Big Eye drifted, he would see his guide star drift, and he would touch buttons on the paddle to nudge the telescope until the crosshairs met again on the guide star. He kept his 1950 model Eveready flashlight in his pocket for emergencies. Prime focus had a padded tractor seat. He kept still on the tractor seat, occasionally glancing up at the sky to note the constellations turning around the North Pole. "You really have the temptation to just stare at the sky," he recalled. "One could imagine space travel is like that." Up in prime focus, it seemed that all he needed from life was a tall ship and a star to steer it by. He wore an electrically heated flight suit, from the Army Air Corps. Prime focus had an intercom speaker that flung Bach cantatas at the stars.

Soon after first light on the Hale, a Palomar gadgeteer by the name of William ("Billy") Baum had sent away in the mail for a huge pile of war-surplus electrical hot suits. The price: one dollar each. Pretty soon just about every astronomer in the United States had to have an electrically heated flight suit. The next time the observatory ordered a batch of hot suits, the price had risen 25 percent—gougers were already charging $1.25 apiece for them. An electrical suit could help neutralize the cold, but it did nothing for the curse of prime focus—the agony of the bladder. Schmidt usually took a break around midnight, but some astronomers did not. When you have been waiting for perhaps a year to get a few nights on the Big Eye, every minute on the telescope can seem valuable; much too valuable for the astronomer to spare a trip downstairs to urinate. This had led to bizarre practices. During recent years some of the electronic cameras in prime focus needed to be cooled with shaved dry ice. The astronomers would carry their dry ice up in a thermos bottle. Several times during the night they would pour ice chips into the camera, to keep the camera cold. Eventually the thermos bottle would be emptied of its chips, and the astronomer would then urinate into the bottle and tighten the cap. It is said that on at least one occasion a groggy astronomer had forgotten what he had done, and thinking his thermos contained ice chips, had poured a steaming thermosful of urine into an expensive scientific instrument at prime focus.

The electrical hot suits deliver one kilowatt of heat at full power, and they are still in use on Palomar Mountain, festooned with duct tape and loose wires, although some of the younger astronomers consider those suits to be nothing more than execution shrouds. What if—heaven forbid—the curse of prime focus overwhelmed you and you pissed inside an electrical hot suit? You could be electrocuted, with a great sizzle and in a cloud of boiling urine. "You could burst into flames in prime focus," Don Schneider remarked, "and you know, in space no one can hear you scream."

Rudolph Minkowski had all kinds of problems in prime focus. His electrical suit was too tight. He almost suffocated trying to put it on, until his wife finally opened it up and sewed a gusset into the paunch. Minkowski also had trouble learning how to use the telescope. Byron Hill, who had supervised construction of the telescope, showed Minkowski all the buttons on the control paddles in prime focus, and then Hill spent the night at the night assistant's control desk (which in those days stood at the foot of the telescope) in order to help Minkowski, in case Minkowski needed help. All evening Hill would hear bangs and groans coming over the intercom. The telescope would keep swaying back and forth.

"Do you need any help?"

"I'm fine," Minkowski growled.

Toward morning, on one occasion, Byron Hill got fed up. The telescope was jerking all over the place. "What in the hell are you doing? Get your elbow off the paddle!" he shouted.

"What of it? I'm doing it on purpose," Minkowski said.

During the next year Minkowski's problems in prime focus deepened. The night assistants could hear him talking to himself, wheezing, grunting. He made a sound like a bear—"Uuuunnh!" The night assistants thought these sounds were so interesting that they made a tape recording of them and passed it around. Byron Hill finally figured out what was going on. The seat in prime focus was a hard wooden platter about half the size of Minkowski's ass. "That

An astronomer at prime focus, sitting in the prime focus cage at the top of the Hale Telescope and staring into the great mirror at light coming out of the deep universe, as imagined by Russell W. Porter in 1940. The astronomer is wearing a tailored and pressed suit, with a pocket handkerchief and wing-tip shoes, and his hair is slicked back with some kind of pomade, maybe Wildroot Cream oil. Such a perfectly dressed astronomer is a purely theoretical being, and has never once worked at the Hale's prime focus, a place known for bitter cold, long nights, pure ecstasy as the Big Eye swings through the stars—and the agony of the bladder. (Photograph courtesy of Palomar/Caltech)

little seat nearly killed Minkowski," Hill recalled. "I couldn't stand to think of the agony up there." One day Hill drove down the mountain to a dealer of farm machinery and paid cash for a tractor seat. Hill said, "I went to a lot of trouble altering that seat to fit Minkowski. I built it up with padding. Then the rat went and took off forty pounds."

Smoking was strictly forbidden in prime focus, but prime focus reeked of cigarettes whenever Minkowski had been up there, although the astronomers could not figure out what he was doing with the butts. The night assistants knew. They said that Minkowski was tossing his butts out of the prime focus cage, where they often fell fifty feet down through the telescope and landed on the mirror. The mirror was a perfect ashtray, because when the telescope moved, any cigarette butts lying on the mirror would just roll into a gutter around the edge of the mirror and out of sight. The night assistants claimed that they had taken fistfuls of butts from around the mirror, which made Walter Baade's hands shake just to think of it. Baade lectured Minkowski about the effects of ashes on Pyrex glass, to no avail. The Prime Focus Spectrograph was a frail scientific instrument, studded with knobs that tormented Minkowski. When he could not get a knob to turn the way he thought it should, he would give it what was called the Minkowski Treatment. First he would wrap a fist around the knob and really twist it. If the knob still refused to turn, Minkowski would utter one or two obscene remarks in German, throw his cigarette overboard, and produce a little pair of pliers from his pocket and absolutely destroy the knob. On one occasion Minkowski could not get the camera on the spectrograph to open up so that he could change photographic plates. The problem was a pair of wing nuts. The wing nuts would not loosen up, even when he worked at them with the pliers. He did not realize that he was turning them the wrong way, and was actually tightening them. The next day the engineering crew found that Minkowski had frozen the wing nuts, and they had to cut them off. They replaced them with clamps having knurled thumb grips, in order to encourage Minkowski to keep his pliers to himself, but, as Byron Hill said, "You could make the telescope astronomer-proof, but you could forget about making it Minkowski-proof." Nevertheless, once Minkowski had gotten settled on the tractor seat with a cigarette and had gotten the colors of a radio galaxy streaming through the diamond dealer's watch fob, Minkowski went into a kind of hibernation. His grunts mellowed into sighs, the Hale Telescope leaned over into the night, and Minkowski and the sky became one. Minkowski is one of the few members of the human species to have a galaxy named for him. Minkowski's Object, a very peculiar galaxy, sits in the constellation Cetus, the Whale.

▪ ▪ ▪

On the night of December 27, 1962, Maarten Schmidt took a two-hour exposure of a bright star that lay next to the little radio streak known as 3C 273. He finished the exposure just before dawn. Breaking apart the camera, he slipped the tiny exposed plate into a lightproof box. He worked quickly, because during the moments when the plate was in the open air, a meteor could flash overhead, exposing the plate and ruining it. The next afternoon he developed his bits of glass in the darkroom. When they had dried, he studied them through a magnifying glass like a jeweler's loupe, while jotting notes on the yellow graph paper that he used for recording his thoughts: "Dec. 27. 3C 273. This is the bright star at the end of the streak. Everything is strongly overexposed."

The star had practically roasted his plate. He noticed that the star was emitting strange colors. It was one of those radio stars: "There is a broad emission line at 3,250 [angstroms, a common measure of wavelength] . . . Also some fine regularly spaced emission lines around 3,400. . . . There must be much more, we need a lighter exposure."

The human mind forever wants to see tiger stripes in the forest. As it would turn out, there were no regular lines in that particular exposure. The plate was grossly overexposed.

Two nights later, on December 29, having been busy with radio galaxies, Schmidt got around to taking another spectrum of the radio star. Looking at it through the eyepiece and tweaking the Hale Telescope, he watched the star move onto the slit. He was really quite surprised at how bright this star was, at least by the standards of the Big Eye. He was used to looking at faint galaxies— galaxies that he could barely see when he stared down on them in the mirror. "You were always worrying as to whether you were seeing ghosts," he would remember. "You spent a long time staring at the field while you were setting the telescope on the object. You had to use averted vision—look away to one side, and then you'd see the object, or maybe you wouldn't see it. One time I spent four and a half hours taking a spectrum of a galaxy. When I developed it, I had an absolutely blank plate. I had totally imagined that galaxy." This 3C 273 was no ghost. "It was outstandingly bright," he recalled. "I could just barely see color in it. Optically it looked rather blue." The following afternoon he saw that he had made a good exposure of it—he saw all kinds of emission lines. He decided that the faint streak next to the star must be some kind of a jet emerging from the star.

He returned to Palomar Mountain at the end of January to work on more radio galaxies. He also tried to take more spectra of 3C 273. On the first night he overexposed another plate. He could not get used to photographing these bright stars. The following night he shot the star in a very brief exposure, and then he wore out the night trying to soak up a spectrum from the fine jet that protruded from the star. At break of day, feeling wobbly and eerie and happy, as he always did after working hard on the sky, Maarten reluctantly turned away from the eyepiece and returned to earth, holding, in his lightproof box, a few bits of glass containing what he believed were images of starlight. When he developed the plates, he saw that the long exposure of the jet had produced absolutely nothing: "Jet—needs further looking into."

He went back to Pasadena. He had taken several plates of 3C 273 by now. The spectrum showed six emission lines. As usual, the lines did not correspond to any known form of matter. He described the lines to colleagues, and nobody could explain them. Meanwhile the British journal *Nature* wanted to publish some articles on these peculiar radio stars. Maarten agreed to write an article.

Down the hall from Maarten's office in the Robinson building at Caltech, Jesse Greenstein had been working on an article for the *Astrophysical Journal*. Jesse believed that he had found the astonishing secret of a radio star called 3C 48, which was nothing less than this: that 3C 48 was a dwarf star glowing with heavy metals, such as curium, neptunium, and plutonium. One day he walked into Maarten's office carrying a bulging manuscript that described his findings. It was forty-one pages long and contained fifteen tables and graphs. "This," Jesse said, "is the best I can make of 3C 48. If you have any remarks, let me know within a week, and then I'll send this off." It hit Maarten's desk with a heavy sound.

"If I see anything funny, I'll let you know," Maarten replied.

On February 5, 1963, Maarten Schmidt got down to business in his office to try to write his article for *Nature*. He placed some leaves of yellow graph paper on his desk (his manuscript paper) and laid out his glass plates of 3C 273. Each plate held a tiny black-and-white stripe, a spectrum. Some of the stripes were only a quarter of an inch long. He had mounted the plates on standard microscope slides with Scotch tape, and now he dusted them lightly with a rather elegantly patterned handkerchief and inserted the slides, one at a time, into a cast-iron microscope. He pulled off his glasses and squinted.

Even in his best spectrum of 3C 273, the features were hard to see. The spectrum clouded the glass like a streak of smoke. The smoke thickened almost imperceptibly, here and there, into broad vertical bands. These bands were the emission lines. Always he worried that he was seeing ghosts. Schmidt could not get over the notion—which had hit him mistakenly, earlier on the mountain— that he was seeing something organized here, something proportional in these lines. The lines fell at decreasing intervals, going from red to blue, as if they were harmonics of an excited atom. He knew also of an invisible infrared line, discovered by J. Beverley Oke, which he could not see in his plates, but he realized that Oke's invisible line would be spaced regularly with the others. So there would be five regular lines, and two other lines that did not seem regular. Sketching on the graph paper, he tried to construct a model of an atom that might emit harmonics of light. What kind of hot gas could glow in harmonics? "So I got a slight bit frustrated," he would recall. "'Look here, it *is* regular, isn't it?' I said to myself, as it were."

To satisfy himself that his lines were regular, he decided to check their spacings against the Balmer series of emission lines from glowing hydrogen—the most regularly spaced set of emission lines known in physics. The Balmer lines were spaced at decreasing intervals. He measured the intervals of the lines in his spectrum, compared them to the Balmer lines, and suddenly he understood. He was *seeing* Balmer lines in the radio star. He was seeing hot, glowing hydrogen in this radio star—except that the colors of hydrogen were pulled far down the spectrum, toward the red end. That would account for five of the lines—all regular. Now, what about the other two lines? If he moved these other two lines back up the scale to normal wavelengths, what would they be? He pulled out his circular slide rule and spun it. Magnesium. Oxygen.

The radio star was made of normal elements. But it was receding from the earth at about 16 percent of the speed of light. This was a Doppler shift. The object was withdrawing in the general expansion of the universe—in the Hubble flow. This wasn't any star—this was an extragalactic object. A 16 percent redshift would place it around two billion light-years away, among galaxies at the limit of the Hale Telescope's imaging power—galaxies so faint that when he glanced at one in the mirror with averted vision, he wondered if he was seeing a will-o'-the-wisp. And this object was so bright it had twice burned up a plate.

Schmidt, feeling unable to comprehend the enormity of his discovery, opened his door to let in a little air. At that moment Jesse Greenstein walked past. Maarten said, "Jesse, would you come in for a moment? I want to tell you something."

Jesse sat down. Maarten told him that he had found an extreme redshift in a radio star.

Jesse's face went pale. Jesse said, "Oh, my goodness!" A flash of something close to horror crossed his mind at that moment. In an instant Jesse saw that his theory of the radio stars was all wrong. He realized that he had *seen* a redshift in 3C 48! But he had rejected it. Instead he had convinced himself that the thing was a tiny star, drenched with curium, neptunium, and plutonium! He said, "We ought to look at 3C 48."

They dug up Jesse's manuscript, and in a few moments Jesse announced that 3C 48 had a redshift of 37 percent. Jesse's face went into an extreme redshift. He had already mailed his paper to the *Astrophysical Journal*.

In Maarten's words: "Our eyes were opened."

3C 48 was departing at more than one-third the speed of light. It would be five billion light-years away—and it was a brilliant point of light! Maarten and Jesse covered a blackboard with calculations. They could not believe what their eyes told them. They groped with chalk for a way to explain these emission lines without resorting to a redshift. They began shouting. The noise brought Bev Oke into Maarten's office. Schmidt and Greenstein challenged Oke to disprove a redshift in these lines. He could not. Jesse telephoned the *Astrophysical Journal* and asked the editors to suppress his paper. ("It was a fascinating paper," Jesse would say to me as he recalled these events years later. "Except that it was wrong.") By 5:30 P.M., the universe had grown too strange to be contemplated without a drink. Jesse suggested that they go back to his house. When the three astronomers showed up looking for liquor, Jesse's wife, Naomi, was flabbergasted. Caltech astronomers never went out drinking on Tuesday nights. "What's happening?" she asked.

Maarten could not sit on the couch. He walked up and down. If these things—which could no longer be called radio stars—were deep in the universe, then the light they shed equaled the thermal burning of an entire galaxy all at once!—yet crammed into a tiny area, as if some force had crushed a hundred billion stars into a pinpoint and ignited it. And the jet! That jet coming out of 3C 273 looked like a blowtorch flame. The jet would be as long as three galaxies. It was horrible—what kind of force in nature could make a jet of gas three times bigger than a galaxy? "We acted so strange," Maarten recalled. "We were shouting." While Maarten seemed almost dismembered from nervousness, Jesse navigated through the wake of "that first terrible afternoon," as he called it, with the help of Schimmelpenninck cigars ("The Apex of the Dutch Cigar Industry") and Chivas Regal Scotch whisky. "Let me have some of that, too, please," Maarten had said, pointing to the whisky. Looking back on it all now, Jesse says, "We had broken through a bubble in which we had been trapped. That is a deep feeling for a scientist. When you are working within a field and a discovery like this happens, the feeling is absolutely incommunicable; it's organic."

Maarten drove home. For many hours that night in his living

room, he has recalled, "I paced up and down like a caged tiger." He was thirty-three years old.

Corrie asked him what was wrong.

He said, "Something terrible happened at the office today." He told her that he had found an object among the most distant galaxies that burned with a terrifying light. He would have to publish. So little time, and what would he say about it? Walking around the living room, he asked himself, "Are you making a mountain out of an antheap? Or if not, what do you say? Boy, I will have to say something!" He wondered if he was overlooking some simple, innocent, rather ordinary explanation for these emission lines. What a fool he would make of himself if he published an article declaring that a bright star was two billion light-years away! "It all came clear, already, what the future held," he would later say. "Because if you see very bright objects with such large redshifts, then somewhat fainter ones must have much bigger redshifts." Maarten and Corrie put the children to bed, but Maarten could not sleep. They had recently bought their first television set. He switched it on and tried to watch a show. He realized that many more of these objects would soon be discovered. They would be fainter, of course, because they would be farther away. The next twenty-five years of his life stood in front of him as a straight road pointing into lookback time, and space opened before him into a gulf that sparkled with remote fires. To search for these things would be to probe down into time, almost into a different universe, and to watch a brutal, inexplicable drama occurring in an alien place. When he saw the test pattern on the television, he went to bed, asking himself, over and over again, Is there a way out?

There was no way out of this universe. Schmidt, Oke, Greenstein, and Matthews (the radio astronomer) scrambled to write a string of papers for *Nature*, which all appeared in a row. Schmidt's paper came first—"3C 273: A Star-Like Object with Large Redshift." It was two pages long. There was nothing much to say, really. Nature had turned out not to be complicated and explainable, but uncompromising, simple, and mysterious. Nature offered alibis to no one. These two pages marked a turning point in the history of astronomy, announcing a new heaven spattered with explosive, eerie phenomena—two pages that were a prologue to two decades of astronomy that would reveal pulsars, accretion disks, black holes, gamma-ray bursters, radio jets, gravitational lenses, and the baroque, inevitable logic of the Big Bang, the moment of creation. When more and more quasars were discovered, and when their deepening redshifts pointed his way into lookback time, Maarten Schmidt came to see that when he had dusted that slide with his handkerchief and stuck it in the microscope, he had accidentally stumbled into a quest known as observational cosmology, whereby one tries to figure out the structure and history of the universe by examining it in a looking glass.

As for Jesse Greenstein, he would soon blow a good deal of money amassing a collection of Japanese Zen paintings. Jesse considered his paintings not exactly a consolation for having let the redshift of the quasars slip through his grasp, but rather a lesson. "I had known that 3C 48 had a redshift," he said. "And I had thrown that notion out. 'This is nonsense,' I said." He collected paintings that illustrated Zen koans, riddles. One of his favorites shows an old poet riding in a boat, watching geese fly across a cloudy, moon-illuminated sky. The poet is looking up, and one can hardly see his eyes. The riddle is: How does the old poet catch the geese? And the answer is: He has already caught them.

Jim Gunn and Don Schneider were working in the data room, trying to prepare 4-shooter to scan. They hammered at keys, while Maarten Schmidt chatted with Juan. Suddenly Jim and Don shouted, "Oh, no!" and ran out of the room.

"Anything wrong?" Maarten called after them.

No reply.

A printer in the data room started spitting paper. It said:

OK

OK

OK

Maarten contemplated it with a smile.

OK

OK

OK

"It's saying okay all the time," Maarten said, "but I don't think it's okay."

OK

OK

OK

Jim and Don raced back into the data room. They pounded keys, trying to soothe 4-shooter.

"I don't understand a blessed thing of what is going on at the moment," Maarten said. "Actually it feels quite normal."

"Now it's okay," Jim said, while Don gathered up a mound of coiled computer paper. Maarten hovered over them. He said, "You are boldly going where no man knows what he is doing."

They persuaded 4-shooter to start scanning, and stars and galaxies drifted up the screens, but the galaxies looked very different this time. Each galaxy threw out a vertical smear, resembling a candle flame. The screen displayed images of galaxies as seen through flat pieces of glass known as diffraction gratings. Such glasses decompose light, as does a prism. Jim had placed the glasses in front of the four cameras inside 4-shooter, so that the light of every object in the field of view passed through the glass on its way into the camera. This technique smashes apart the light of everything in the telescope's field of view. The galaxies appeared to be on fire. Each flame was a spectrum emerging vertically from the galaxy. During two nights of scanning, 4-shooter would pass over the same strand of sky twice, taking direct images on the first night, taking images through diffraction gratings on the second night, thereby acquiring pictures of about 120,000 objects along with their broken colors, all of which information would be recorded on tape. Don's computer would later combine the images to intensify the light, and then automatically search the spectra for emission lines typical of quasars. In that way the team hoped to find quasars.

"We are working!" cried Maarten. He strode to the stereo, and within moments he had found Mozart.

The astronomers gathered around a screen to read the spectra.

"Maarten, look at this," said Jim, his finger tracing bumps and gaps in a candle flame. "That's an early M star." (He later explained to me that an M star is a cool, reddish, aging star.)

Maarten took off his glasses and squinted at it. "Ja," he said. "A rather blue M-type."

The sky on the television screen was a mass of blots and smears. Some spectra had dark cuts in them—absorption lines. Others showed swellings—emission lines. The astronomers noticed many M stars. M stars, they said, superficially resembled quasars. "Most of them are pretty close to us," Schneider said. "Within a few thousand light-years away." Later he touched the screen. "There's an emission galaxy," he said, indicating a bright, violent galaxy with something nasty burning in its core, perhaps a mini-quasar.

"Ah, yes," Maarten said, tracing the spectrum with his finger. "Look at that. An N galaxy." He pointed to a bundle of horizontal spikes in the candle flame. "You can see emission lines, but it is clearly a galaxy and not a quasar," he said.

These multiple transits on the Big Eye suggested the rhythm of long-distance driving across North America at night. Galaxies sparkled on the video screens like the lights of lonely towns. The talk rose and fell, and often the astronomers stared in silence.

"A carbon line?" Don said, touching another spectrum moving on the screen. "This could be a quasar."

"For this we need the supercomputer," Maarten said. He twirled his circular slide rule. "The emission lines on that object are a bit too far apart to be carbon. So I'd say that's a magnesium break. It's only an emission galaxy. Sorry, gentlemen."

Later Maarten said to Jim, "I think it's about time we saw a quasar go by, James, don't you?"

"Absolutely."

"We need to see a quasar," Don said.

▪ ▪ ▪

One night the door of the data room swung open, and an astronomer who had been working on the forty-eight-inch Schmidt telescope walked in. "I am really ticked off," he said. "It's the second bomber tonight."

"The second what?" Don asked.

"B-52! Some idiot up there crossed right through my field with a blinking strobe light and all kinds of insanity. He just ruined my plate. I think those guys are vectoring off our domes on night bombing runs." He leaned over and looked at the screen. "Hey. This is impressive. Have you seen any quasars?"

"We would surely like to," Gunn said.

"Gunn, this is really impressive. You could sell tickets to this."

Gunn called to Schmidt, who was on the other side of the room, "Maarten, did you hear that? Who cares about a few spectra when you can get a hundred thousand?"

"Ja, that's pretty good!"

On another occasion Schmidt, who had been walking restlessly around the room, suddenly whirled on his feet. He had seen something moving on the screen, out of the corner of his eye. "By God," he said. He ripped off his glasses. He grabbed a ruler and put it against a spectrum floating up the screen. He pulled out his circular slide rule and twirled it. "Redshift, let me see! Yes! That was a quasar!"

The astronomers' chairs, which were on wheels, thundered up to the video screen. "That was a bright one," Gunn said as the quasar disappeared at the top of the screen.

During the past year, Don Schneider had been writing a massive engine of software to find quasars, which, everyone hoped, would find this quasar again. To my eyes the quasar had been indistinguishable from the hordes of spectra splattered across the monitor. Maarten's reaction reminded me of a fly fisherman, working a slick in a Maine river, who hears an odd, faint skitter of droplets, turns, and, without missing a beat in his cast, drops the fly three feet up-current from the boil of a resting salmon.

A blast of white flooded the screen. "A little tiny star," Jim mused. "You couldn't see it with the naked eye."

Don remarked, "I just hope we don't cross Orion's Belt."

"Say! Which way were we supposed to be going?" Maarten joked.

"Or the moon," said Juan Carrasco. "Then it would be good-bye screen."

A cluster of galaxies drifted up the screen, spectra bleeding out of them.

"It looks like those galaxies multiplied," said Juan.

"It's the other way around," Don said. "They eat each other up."

"Is that right?" said Juan.

Maarten pointed to the screen. "That little galaxy, Juan, is that big one's lunch for the next billion years."

▪ ▪ ▪

"Something terrible happened at the office today." After Christmas, 1963, at a conference held in Dallas, astronomers argued about the proper name for these quasi-stellar objects. Someone proposed the name Dallas Stars. Someone else thought they should be called Kennedies, in memory of President John F. Kennedy, who had been shot in Dallas the month before. (The term *quasar* finally became official in 1970.) Maarten Schmidt began receiving record amounts of dark time on the Hale Telescope. Sitting alone in prime focus and zeroing the crosshairs to the music of Bach, Maarten Schmidt broke open the universe.

On April Fool's Day, 1965, the *Astrophysical Journal* stopped the presses to wait for a letter to the editor from Maarten Schmidt. The letter was no joke. After two years of work he had found five quasars. By a series of tight deductions, he had linked the emission lines of the five quasars into a ladder of logic that had taken him into breathtaking distances. One quasar, he found, was redshifted by 70 percent. Expressed as a ratio, that quasar's redshift was 0.7. (Astronomers usually express redshift as a ratio rather than as a percentage.) Another quasar had a redshift of 1.03—redshifted by 103 percent. The monster was a quasar called 3C 9, with a redshift of 2.01—an awesome 201 percent—and yet this quasar was bright *blue* in color, because a normally invisible ultraviolet glow in the spectrum, called the Lyman alpha line, had been redshifted down into visible wavelengths, where it tinted the quasar with the color of a pale sapphire.

Astronomers do not know the exact distance from our neck of the woods to a deeply redshifted quasar, because astronomers have not yet been able to link redshift to a distance scale. For example, the quasar 3C 9, with a redshift of 2.01, is probably somewhere between ten and sixteen billion light-years distant from the Milky Way; the photons coming from that quasar are anywhere from two to three and a half times as old as the earth. Maarten once said that he was quite proud of his April Fool's quasars. By showing that one could find a quasar with a redshift of 2.01, he had tripled the range of the Hale Telescope in all directions from the earth. He had opened up a shell of explorable space that, in terms of volume, was fifty times bigger than the volume of space that had been available to the Hale Telescope before. In a letter to an editor he had enlarged the known volume of the universe fifty times over. "That," he said, "was the most difficult job I have ever done."

▪ ▪ ▪

The news of quasars reached the pages of *Reader's Digest* in the year 1966, and a copy of that magazine traveled to an old lady who lived on a farm near Heartwell, Nebraska, which was then a hamlet without a paved street, situated on the high, treeless plains south of the Platte River. Her name was Mrs. Gertrude Schneider. On Sunday afternoons her eleven-year-old grandson, Donnie, used to drop by to visit with her and, incidentally, to read her *Reader's Digest*. It was at his grandmother's farm, reading *Reader's Digest*, that Don Schneider first learned about Maarten Schmidt. He began to feel the presence of quasars over his head. "That was when I gave up dinosaurs for astronomy," Don says. "It was the last career switch I ever made. By the time I was in sixth grade, I knew that I was going to be an astronomer, at least as much as a child can ever know what he will become." Don always felt that he had chosen a normal career. "How anyone can go out at night and look up and *not* want to be an astronomer is beyond me."

When he told his parents that he was going to be an astronomer, they were pleased, because they originally had not had high hopes for Donnie. He had been a slow baby. At the age of three and a half Donnie still had not learned how to talk. They had begun to fear that he was mentally retarded, and they had made plans to send him to a doctor. Then one day Donnie was sitting in his grandmother's lap while she read a book to him, hoping to encourage him to speak. She pointed to a picture of a barrel and said, "Keg. Can you say that? Keg."

"Barrel," Donnie said.

"What?"

"Barrel," he said, pointing to the word.

"Are you reading this?"

"Yes."

She flipped the pages, pointing to other words, and he read the words aloud. He had not had anything to say until he could read it first, at three and a half.

Donnie puzzled his mother too. Eileen Schneider taught catechism in Sunday school, and when Donnie was in first grade, he was one of her pupils. She tried to explain to the children what would happen at the end of the world, at the coming of the Son of Man. She read from Matthew: "'The sun shall be darkened and the moon shall not give her light and the stars shall fall from heaven and the powers of heaven shall be moved. . . .'"

Donnie's hand shot up. "Mother," he said, "the moon doesn't give light. It reflects the light of the sun."

That did not make catechism any easier.

His father, Donnie Ray Schneider, was a sharecropper and the mayor of Heartwell. He farmed corn, wheat, and sorghum on sections of land around Heartwell that belonged to other people. He worked as hard as any man in Heartwell, and people there say that Donnie Ray Schneider never walked anywhere, he ran. He wanted to save up enough money to buy a farm of his own, but there were times when the crops would fail two years in a row, and then he could barely feed his family, although he continued to buy books for his oldest boy, because Donnie was a reader. Donnie Ray figured that someday Donnie would work alongside him and help him build a farm, until the boy announced that he was going to be an astronomer, which was really all right, since it had already become clear that he was not going to make much of a farmer. Donnie Ray had to carry the boy out of bed most mornings and put him on the tractor and wrap his hands around the steering wheel and turn the ignition key before the boy woke up, which perhaps explains what happened when he left Donnie alone one day on the tractor, pulling a giant disc and harrow. Donnie threw the tractor into gear and began to disc the land. He enjoyed open plains, deep sky, and horizons. The sky was always a presence in Nebraska, where one's eye could jump to the edge of the world without hitting so much as a tree. The tractor churned along, and then it came time to refuel.

He had never refueled the tractor while pulling a large piece of machinery. It was a tricky job. He had to drive the tractor alongside a pickup truck. The pickup held a tankful of diesel fuel and a hose. He made a slow pass alongside the pickup truck—a practice flyby. He was too far away, decided to circle around again, and gunned the throttle. As Don tells the story, "I heard this tremendous crash. I looked back. All I could see was a cloud of dust and a dancing pickup truck." He had forgotten about that disc and harrow. They had caught the pickup truck and were discing and harrowing it. "The disc," Don says, "didn't even know the pickup was there." Don unhooked the tractor from the wreckage and drove the tractor like mad to another field, where his father was cutting wheat. "Dad," he said, "I think I destroyed your pickup truck."

The words of the mayor of Heartwell were: "Okay, Donnie, we've gotta make hay while the sun shines. I'll take a look at it later."

Wheat harvest arrived in July. At the same time they had to begin irrigating the corn. Don would get up with his father before sunrise, load a quarter of a mile of irrigation pipe into a trailer, and then set the pipe in another field, laying the pipe sections by hand. After that his father and his uncle would join forces to start cutting the wheat with combines, and Don would help them. Don's mother or his aunt would make up a supper in a picnic basket. In the hot evening the men, Don, and his cousins would sit in the shade of a truck and eat while the sun went down, and then they would work into the night with the combines, shining lights into the wheat until the dew rose and they had to stop.

Don's father finally saved enough to buy a farm of his own. In 1973, the Schneider family moved outside Heartwell to a yellow house surrounded by a picket of evergreen trees, to keep out the high plains wind, but entropy had a way of coming through windbreaks. Don can remember his father picking up a full thirty-gallon drum of oil and putting it into a trailer in a manner that suggested that the barrel contained popcorn. But secretly at first, and then gradually happening in a more apparent way, Donnie Ray's heart changed into that of an old man. He died of heart failure one April in a hospital in Lincoln, just before planting time. Don was in his second year at the University of Nebraska. The other kids in the family were too young to work machinery, and his mother had never even learned how to drive a car. Don thought that he would have to plant the crops alone, until a crowd of tractors showed up in front of the Schneider farm. Most of the farmers in Heartwell put in the Schneider crops, after which Don and a hired man took over. "That summer," Don recalls, "I just did what the hired man told me." Don's mother learned how to drive a car. By autumn, Eileen Schneider felt that she faced a decision: either her son became a farmer, as his father had been, or he got a college education. Without consulting Don, she auctioned off all of the family's farm machinery in order to make sure that Don finished college.

She has remarried and now lives modestly but comfortably on the farm, renting the land to other farmers.

Don graduated from the University of Nebraska in 1976, and went to graduate school in astronomy at Caltech. There he stepped into a class taught by Jim Gunn. The subject was cosmology. Gunn had a habit of saying in class, "As you learned at your mother's knee," and, turning to the blackboard, Gunn would produce a gigantic mathematical expression describing the subtle curvature of spacetime. Don had learned catechism at his mother's knee, but this was pretty good stuff too. He eventually wrote his Ph.D. thesis on cannibal galaxies. His prime suspect for cannibalism was a nightmarish object—a pack of nine galaxies in a feeding frenzy, interdevouring one another. He concluded that they would merge into a glob—into a supergiant galaxy—almost immediately, in a couple of billion years. Maarten Schmidt then hired him as a postdoctoral fellow, to help Schmidt study quasars. When Don's fellowship ran out, he and Schmidt continued to work as collaborators, while Don moved to the Institute for Advanced Study, in Princeton, New Jersey, where he prepared his image-processing program to handle pictures taken by the Hubble Space Telescope.

Don lived in an apartment next to the Institute, which he decorated simply. He put down carpets of computer paper by the front door to keep people from tracking mud around the living room. On the wall he hung some needlework that his sister had made for him, and a small, unobvious crucifix. He stocked a bookcase with novels by Charles Dickens, Anthony Trollope, Jane Austen, and Mrs. Gaskell. He drove an ancient Chevy Nova with a vinyl roof. The car had almost no mileage on it because he walked everywhere he could. He walked to the other side of town to go to Mass and walked back again. His eyes were startlingly blue, as if they had absorbed ultraviolet light from high plains skies. He arranged his days with an almost Franciscan clarity, giving fifty to ninety hours a week to galaxies and quasars for a wage that was manifestly not upscale. That he had gotten off a tractor in order to become an astronomer he regarded as fate, since he had wound up back on a tractor seat at prime focus on the Hale Telescope. One day, as it so happened, he was in Nebraska, walking through

186 ■ *Richard Preston*

a muddy field. It was autumn. Dark clouds were folding and rolling along the horizon, and a strange wind was blowing, which seemed to originate not from the earth. He had lost the Hale Telescope somewhere, which filled him with a sense of terrible loss. There was a white barn in the distance. The wind tugged and pulled at him. He kept walking. He arrived at the barn. He pulled open the barn doors, and there was the Hale Telescope. Then he woke up in his apartment in Princeton. The Hale had entered his dreams.

During cloudy weather on Palomar Mountain, he finished chapters of novels by Anthony Trollope quicker than you can split an Oreo. Maarten Schmidt said to him once, "You belong in the forties and fifties, Don. And I don't mean of this century." Don had fallen in love with Elizabeth Bennet, the heroine of *Pride and Prejudice*. In a thoughtful frame of mind one evening, standing on the catwalk and watching the stars come out, he remarked to me that he had recently celebrated his thirtieth birthday. And, quoting from Jane Austen, he added, "It is a truth universally acknowledged that a single man in possession of a good fortune must be in want of a wife." Nevertheless, he had begun to suspect that the sky might more easily yield deep quasars than the earth a suitable wife. So he contemplated the last words of one of his favorite books, *The Count of Monte Cristo*, which were, "All of human knowledge is summed up in these two words: 'wait and hope.'"

■ ■ ■

While Don Schneider was in junior high school destroying farm machinery, Jim Gunn was getting lessons in hand-to-hand combat. The Vietnam War peaked around the time Gunn finished graduate school. "I have always been something of a physical coward," Gunn once remarked, evidently thinking that he was giving a plausible explanation for why he signed up for paratrooper school at Fort Benning, Georgia. "Boot camp was pure shit," he recalled. "But the jumping was great." Great, that is, until the Caltech astronomer Jesse Greenstein heard about the jumping. Jesse got on the telephone. With whom Jesse talked is not known, but he carried influence in the government, and it can be surmised that his words reached the Pentagon. Higher powers sent Gunn back to Califor-

187 · First Light

nia, where he wound up a captain in the Army Corps of Engineers, doing research at the Jet Propulsion Laboratory.

After being decommissioned from the Army, Gunn went to Princeton University, where he quickly became known as a theorist. In 1970, Gunn returned to Caltech, where he extended his reputation as both a thinker and a skilled observer, while at the same time his hands began to accumulate small scars. He built an electronic camera for the Hale Telescope that used a night-vision tube as a sensor. He built another. Then he built a computerized spectrograph for the Hale Telescope, working with Bev Oke. He built a house for himself and Rosemary. He built quite a lot of the furniture in the house.

There is a joke among astronomers that goes like this. Question: What is the difference between a theorist and God? Answer: God has only one explanation for everything. As a theorist, Gunn has provided our species with a fair number of explanations for things taking place out there—not that it does us any obvious good to know, but it is nice to get the news. At the heart of the Gunnish view is the idea that galaxies are continually forming and dying, feeding or losing mass, taking in material and vomiting it out. Gunn felt that the Milky Way was still growing, still sucking in material from intergalactic space around it. Gunn felt that the universe was a dynamic system, tending toward involved shapes. "Astrophysics is engineering," Gunn said. "You try to engineer a galaxy, using some rules of thumb. Nature is almost always more complicated than you think, but if you bang your head against simple models for long enough, you can sometimes figure out what nature is doing." A galaxy to Gunn was a gadget; and it pleased him to model a universe that worked.

Around Caltech, where there never seemed to be enough money to pay for chic equipment, Gunn applied his wits to other problems. He and Roger Griffin, an English astronomer, used parts from a child's toy to build what is known as the Radial Velocity Machine. Griffin bought a set of Meccano motors. (A Meccano set is the English version of an Erector set.) "The motors were perfect," Gunn said. "They were plastic and they had gearboxes." Griffin wired up a dozen Meccano motors, and Gunn soldered some circuit boards.

They attached the motors to an aluminum frame, added some bolts, some tape, and a Plexiglas lens polished with Brasso metal polish, and what came out was the Velocity Machine, which they installed on the Hale Telescope. It is still in use on the Hale, where it accurately measures the velocities of faint stars whipping in and out of the Hyades, a cluster of golden stars in the constellation Taurus.

On August 9, 1974, Richard M. Nixon resigned from the presidency of the United States. A few weeks later, a United States Congress, nervous about inflation, budget cuts, and the future in general, voted to allow the National Aeronautics and Space Administration—NASA—to spend three million dollars to look into the possibility of building a large space telescope. The vote had been a close call. It brought megascience to the world of astronomy—big budgets, big politics, and big bureaucracies. Thirteen years later the Hubble Space Telescope was sitting in Sunnyvale, California, in a clean room belonging to the Lockheed Missiles and Space Company, awaiting launch in the space shuttle *Atlantis.* (It was eventually launched in 1990.) The Hubble Space Telescope is a complete orbiting observatory. It has a total of five instruments on board. It is four stories tall and weights 25,500 pounds. It cost, ultimately, $2.5 billion, or $6,000 an ounce. The Space Telescope is worth something like sixteen times its weight in gold. The Space Telescope's mirror is only ninety-four inches across, modest by today's standards, but since the Space Telescope orbits above the atmosphere, it is seeing things never dreamed of. A pie-shaped box the size of a grand piano is plugged into one side of the Space Telescope. This is the Space Telescope's main imaging camera, the Wide Field/Planetary Camera. It cost $60 million. The story of how it got there involves Jim Gunn.

NASA moved slowly into the Space Telescope project at first. A major technical problem turned out to be the lack of an effective sensor for the Space Telescope's main imaging camera. A group at Princeton University had been experimenting with vacuum-tube sensors—these were similar to night-vision tubes—but the tubes tended to break down. Tubes did not seem reliable enough for the Space Telescope, and after all the trouble with Congress, NASA did not want to ask for a billion dollars for the Space Telescope until a workable camera could be put on the drawing boards.

Enter James A. Westphal.

Westphal is an astronomer who grew up in Tulsa, Oklahoma, where his father ran an auto-repair shop. When Westphal gradua-ted from high school, he got a job as a "jug hustler" for a seismic crew exploring the Oklahoma panhandle for oil and gas. The jug hustler plants a string of geophones, which pick up sonic vibrations from explosives, and Westphal's pay was thirty-five cents an hour; for overtime, he got *half* pay. After a few promotions Westphal managed to save enough money to go to the University of Tulsa, where he picked up mostly B's, along with a bachelor's degree in 1954. He went back into the oil industry, where he got a job running a well-logging crew in Mexico. Then he got a job with Sinclair Oil, in Tulsa, to investigate what he calls "unorthodox" methods for finding oil. Westphal pointed sensors at the ground, hoping to pick up gamma rays coming from oil. He pulsed radio waves into the ground, hoping to get an echo from something greasy. With a giant computer Westphal analyzed gravitational warps that might hint of petroleum. Westphal passed word around Oklahoma, Texas, and Louisiana that Sinclair Oil wanted to talk to any person or persons who knew or believed they knew something about finding oil—and diviners and psychics showed up at West-phal's lab, to demonstrate their witching wands and their bottles filled with vitamins and hung on strings, bottles that were supposed to wiggle when dangled over oil. The management of Sinclair Oil had little idea of what Jim Westphal was up to, which was all right by him.

Shortly after the Soviets put up the *Sputnik* satellite, Sinclair's management began to wonder if there might be some business opportunities in outer space. Management invited Westphal to a meeting of the board of directors in New York City, to tell them whether Sinclair ought to be in outer space and, incidentally, to tell them what he was doing out there in his lab in Tulsa. After advising the directors to avoid outer space, Westphal placed a small black box on the table. He said that recently his Tulsa lab had had a breakthrough. He would like now to share it with the board of directors. He said that the box wasn't dangerous, although it could produce mental effects. The atmosphere in the boardroom became hushed. He said, "This is a model of a manager." He hit a switch on the box. There was a whir. The box's lid opened. A hand shot out, seized the switch, turned it off, and went back into the box. Westphal said, "As you can see, you put in some input to a manager, and in a little while, why, that's what happens to your input." The hush turned into total silence, which was suddenly broken by a choking sound—one man laughing hysterically—the comptroller of the corporation. (The device was first imagined by Claude Shannon, of AT&T Bell Laboratories, who also invented information theory. Westphal heard about Shannon's idea through the grapevine, and Westphal was probably the first person to actually build the device. Westphal's Model of a Manager became a popular item in joke shops, although Westphal never made a dime from it.) "We were a bunch of free spirits," Westphal once said, "and we torqued the system and brought Sinclair Oil kicking and screaming into the new world."

After doing that, Westphal landed at Caltech as a laboratory technician. He became, in effect, a jug hustler for Caltech. He proceeded to build a high-pressure aquarium tank capable of holding living marine organisms brought up from deep-sea trenches. He took photographs of the moon and of coral reefs. He became interested in volcanoes and gold mirrors and infrared stars and galloping glaciers in Alaska and the atmosphere of Venus. His problem at Caltech, as he described it, was that "I was never able to figure out what I was supposed to do around here." Westphal began fooling around with night-vision tubes, turning them into cameras for the Hale Telescope. He and an astronomer named Jerry Kristian built a camera containing a Silicon Intensified Target vidicon—a high-voltage night-vision tube. They took the camera up to prime focus. Prime focus is meant to hold one person. With the help of ten or twenty yards of Palomar Glue, Westphal and Kristian installed their night-vision tube in prime focus, along with a computer, a tape recorder, an oscilloscope, a television monitor, a ten-thousand-volt power supply, a tangle of power cables, and both of themselves. They aimed the Hale Telescope at the Milky Way, and then, just for the hell of it, they charged the night-vision tube with ten thousand volts of electricity—maximum redline power—and let first light fall into the tube. A bunch of stars came up on the television screen. Kristian, peering at a star-finder chart, began to complain. He said that the telescope must be pointing poorly. He said he could not recognize any of the stars. Then Kristian began to scream, and that was when they realized they were seeing stars that had never appeared on any star chart. They had warped the Hale into hyperspace and had zoomed into an unknown neck of the galaxy. Westphal let out an Oklahoma wildcatter's "Yaa-hooooooooo!" and Kristian laughed until he gasped—"I almost peed my pants," he said. (With all those high-voltage wires around, that could have killed them both.) Westphal and Kristian's camera is now in Washington, D.C., in collections of the National Air and Space Museum. Jim Westphal had become a Palomar gadgeteer.

Meanwhile three engineers at the Jet Propulsion Laboratory in Pasadena—their names are Gerald Smith, Frederick Landauer, and James Janesick—had been investigating CCD sensor chips for use on the Galileo space probe, an unmanned NASA spacecraft then planned for a voyage to Jupiter. Westphal, talking with these engineers, came to understand that a CCD chip would be a good sensor for a ground-based telescope. He passed that rumor along to Jim Gunn. "It dawned on us," as Westphal tells it, "that if we could improve these chips a little bit, they could just wipe out everything else in ground-based astronomy." The engineers were saying that a chip could be made one hundred times more sensitive to light than any photographic film. To hang a supercharger like that on the Hale Telescope would be exactly equivalent to building a telescope with a mirror one hundred times larger than the Hale's, only much cheaper.

On October 19, 1976, Jim Westphal attended an all-day meeting at Caltech, as an onlooker. The main participants were the members of the Space Telescope's Phase B group—a NASA team in charge of the early design of the Space Telescope. The question at hand was critical and immediate: What in the world should be done about a main camera for the Space Telescope? Smith and Landauer, from the Jet Propulsion Lab, stood up and talked about CCD chips. Westphal presented some lore on CCDs. Robert O'Dell, the group's chief scientist, drew a picture on a blackboard of a mirrored pyramid that could be used to break a beam of light into four parts and bounce them into an array of CCD sensors.

O'Dell said, "You guys think about that." By the afternoon, the group had thought about that and had voted to advertise an open competition, hoping to encourage somebody, somewhere, to surface and solve the problem of the main camera on the Space Telescope. A couple of days later, as Westphal tells the story, Jim Gunn walked into Westphal's office and said, "Jim, we have got to build the camera for the Space Telescope."

"What?" Westphal said. "Buzz off, Gunn."

Gunn said that he was not joking.

Westphal started to feel nervous. He said, "No way, Jim! That's not our style."

"If we don't build the camera, Jim, Caltech will be out of business in astronomy," Gunn said.

"So who do you want as Principal Investigator?" Westphal asked, feeling a weird sensation in the pit of his stomach.

"I want you to be P.I.," Gunn said.

"What? Hell, no, Jim! You can cram it," he said.

Gunn pleaded with Westphal.

"There's just no way," Westphal said.

Gunn insisted that there was a way.

At that point Westphal understood that what he, Westphal, needed was a way out. So he said that he would consider the notion—provided that Gunn would agree *in advance* to be the Deputy Principal Investigator. He felt sure Gunn would not agree to that.

Gunn agreed.

"I have been had," Westphal said. Then Westphal had an idea. He said to Gunn, "Let's each put three or four names on the blackboard. If we can get half of these people to join us, Jim, why, then I'll do it." ("That was my way out," he would later admit. "Right quick I could think of several people who I just *knew* would refuse to get into this thing with us.")

During the following weekend Gunn and Westphal got on the telephone. Westphal called up Jerry Kristian. "Yes," Kristian said. Westphal called someone else—he can't remember now who it was, but he remembers hearing the unfortunate word, "Yes."

"I figured I would have to call Roger Lynds, over at Kitt Peak," Westphal remembered. "I just knew he would bail me out. He said, 'Hell, yes.' At which point I knew in my gut I had had it."

"I twisted Westphal's arm," Gunn would admit. Gunn believed that only Jim Westphal, who had been torquing bureaucracies for most of his life, could successfully build the camera. Gunn feared megascience. He had been secretly terrified that Westphal or somebody at NASA might nominate him, Jim Gunn, to drag the Space Telescope kicking and screaming into the new world. "When I told Westphal that he had to lead the project," Gunn said, "I was merely trying to protect my tail." The team, which eventually grew to twelve astronomers, began to meet regularly, throwing around ideas for how to build a CCD camera for the Space Telescope. They decided to call it the Wide Field/Planetary Camera, or the Wiffpick for short (the word comes from the acronym WF/PC). The Wiffpick would contain a reflective pyramid that would break up a shaft of starlight coming from the telescope's main mirror and bounce it into an array of CCD cameras. It would operate in two modes: a wide-field mode, for making images of faint galaxies, and a planetary mode, a telephoto action for zooming in on planets. (It is interesting to note that Gunn's first good amateur telescope, the one he built in high school, featured a Wide Field and Planetary Camera—as Gunn called it even then.) According to Westphal, "In a technical and a scientific sense, Jim Gunn is the brain behind the Wiffpick."

"Absolutely not," Gunn said. "Westphal is the brain."

In any case, the Wiffpick team saw that it would not be easy to convince NASA that the Space Telescope needed a Wiffpick. One way would be to build a prototype CCD camera, to demonstrate that the idea would work. So Westphal went to a restaurant supply store down the street from Caltech and bought a spaghetti pot for eight dollars. He built a CCD camera inside the spaghetti pot. He took the pot up to the Hale Telescope and fastened it to the Hale with a handful of bolts and a lash-up of Palomar Glue. He let first light into the spaghetti pot. Describing what happened next, Westphal said, "There is just nothing like taking a five-minute exposure and seeing deeper into the universe than anybody ever has before." In the summer of 1977, his team submitted a design for a Wiffpick to NASA. Competing against two other designs, the Wiffpick won.

Jim Westphal has a white beard and a crew cut. His team sometimes refers to him as Captain Fuzzy, although never to his face.

His hands, I have noticed, are usually into something. One day I found Westphal in a lab underneath Caltech with his hands inside an antique (circa 1975) computer tape drive. Anybody but Westphal would have thrown it out years ago. "Not having the tape on the capstan will get you every time," he said as a loop of tape spat out of the drive and circulated around Westphal's hands. "It was a brave act for NASA to select us," he said. "See, legally, I'm an amateur astronomer." He still does not have a Ph.D.—no paper credential other than his bachelor's degree from Tulsa. The Wiffpick was assembled by aerospace engineers in pressurized clean rooms at the Jet Propulsion Laboratory, working under contract with NASA, and neither Jim Westphal nor Jim Gunn touched the camera with a soldering iron.

Nobody at NASA wanted any roaming, septic gadgeteers rooting hog inside the Wide Field/Planetary Camera, even if those gadgeteers had dreamed up the camera in the first place. According to Westphal, NASA has traditionally discouraged the Principal Investigator of a space experiment from trying to engineer the spacecraft. "You don't let people like Jim and I build flight hardware," Westphal said. "It might vibrate apart on launch." The Principal Investigator is supposed to plan the scientific experiments, but trained engineers, hired by NASA, are supposed to build hardware that will not vibrate apart on launch. In effect, the engineers told Westphal and his crew to give them the basic idea for the camera and let them worry about the details. One does not tell a gadgeteer not to worry about details.

At one point Westphal hired an optician named Art Vaughan to do a preliminary design of the mirrors and lenses that would go inside the Wiffpick. Vaughan sat down at his kitchen table with a pocket calculator and invented all of the glass for the Wiffpick. He gave his plans to Westphal, along with a bill for a couple of weekends' worth of work. Westphal sent the plans over to the Jet Propulsion Lab, and then the trouble started. The engineers at the Jet Propulsion Lab took one look at those drawings, panicked, and sent them to a major corporation for evaluation. The major corporation sent back a bill in the neighborhood of forty thousand dollars and an answer that Westphal characterizes as, "Yeah, it'll work great." Westphal blew up. "I went absolutely nonlinear," he said. The resulting uproar caused the aerospace people to nickname the Wiffpick team the Woof Pack, apparently because the Woof Pack had snarled at them and slobbered all over them and bit their ankles.

"Those guys wanted to take a picture of the beginning of the universe," one engineer complained.

"It has not been a real smooth interaction," Westphal put it.

"Those Calteckers can be a pain in the ass."

"Catastrophic differences in philosophy."

"We aerospace types need those people in left field or we'd go rolling downhill. But would I try to fly something Gunn or Westphal had built? Sure. I would take it up in the space shuttle and launch it at the sun."

As a part of his blood pact with NASA, Westphal demanded the right to review and authorize any large requests for money by the engineers. "I followed the Golden Rule," Westphal said. "Those what's got the gold makes the rules." Westphal got his hands on the gold—he signed for every major expense. "I could turn the money on and off," he said. "I could stand up in places where I shouldn't stand up and say, 'This is a bunch of crap.'"

NASA officials were generally delighted with Westphal—he showed awesome engineering and scientific capabilities, and he saved plenty of taxpayer dollars—but one day Westphal went too far. He stood up at a big NASA meeting, and in front of a large number of officials he fingered a fine point of the Space Telescope's design as unworkable crap. An outraged NASA official telephoned Marvin ("Murph") Goldberger, the president of Caltech. The NASA man told Goldberger that NASA was having trouble with a loose cannon. He said to Goldberger, "We want to get Westphal under control. We want him *fixed*. Tell us how to fix him." Murph Goldberger allegedly replied, "I would like to know that myself. If you ever figure out how to control Jim Westphal, please let me know right away."

▪ ▪ ▪

One of the key figures in the development of CCDs for astronomy is James Janesick, a scientist at the Jet Propulsion Lab. He is known as one of the Jims—the other two being Gunn and Westphal. Jim

Janesick, a handsome, lanky person with a self-deprecating manner, had dropped out of high school in the sixties to play lead guitar for a band called the Tangents. The Tangents developed a sound similar to the Rascals. They cut a few singles and made some television appearances. For a short while the Tangents were more popular than the Rascals. Their single "Good Times" brought them onto the charts in Detroit. It looked as though the Tangents were about to take off—and then the Rascals took off. ("They were better musicians," Janesick thought.) Janesick eventually wound up working as a technician for the Navy. One day he saw a classified ad in a newspaper, saying that the Jet Propulsion Lab was looking for an expert in CCD cameras. "I didn't know what a CCD was," Janesick recalled, "but the ad said that they were an equal-opportunity employer." He got the job, whereupon he became entranced with a silicon crystal that could seize a few stray photons that had come from a galaxy far away and long ago. He performed experiments on CCDs—flashed ultraviolet light on them, vaporized gold and platinum on them— and thus discovered ways to make them more sensitive to light. Nobody fully understands how a CCD works, even though it is possible to build one. Janesick said, "Trying to understand what Mother Nature is doing inside the chip is like peeling an onion. Nature allows you to go as deep into the chip as you want, and you will always be entertained." Janesick set about building a signal processor—an elaborate amplifier—that would pluck individual electrons from a CCD and turn them into a crisp image of a galaxy. Janesick thought he knew something about amplifiers.

So did Jim Gunn, who had started building amplifiers in high school. These two Jims got into a serious amplifier contest, to see whose amplifier would fly inside the Wide Field/Planetary Camera. Janesick would build an amplifier, Gunn would build another, and Janesick would build one better. The quality of a CCD signal processor is determined by how many extra electrons of "noise" it introduces into the faint signal coming out of the CCD chip. The fewer noisy electrons, the cleaner and better the amplifier. "One electron may not sound like much," Janesick said. "But in this business, one electron is worth a billion light-years." Janesick's circuits were powerful, elegant, and simple. Gunn's were virtually indecipherable. "Gunn's circuits were always better than mine by two electrons," Janesick said. Gunn said, "If I can make something better by making it more complicated, then I will bloody well make it more complicated." Janesick's signal processor will fly inside the Wide Field/Planetary Camera and Gunn's will not, because, as Gunn mildly put it, "My circuits might not survive the launch."

"In our naïveté or stupidity, Gunn and I probably think that we could build one of those cameras for as little as five million dollars," Westphal said. "But NASA would refuse to fly it." No nuts or washers shall come flying loose in zero gravity. No spaghetti pots shall be lashed to the Space Telescope. The Wiffpickers, as a team, were given about three hundred hours of observing time on the Space Telescope as their reward for designing the Wiffpick. The Wiffpickers had laid their plans, and their plans were imperial. They would point the Space Telescope into the hearts of quasars. They would try to find supermassive black holes. They would drill wildcat wells into lookback time, hoping to pull up images of galaxies in the tormentaria of birth, early in cosmic time. They would try to detect earthlike planets orbiting nearby stars. Of the Wide Field/Planetary Camera, Westphal said, "I just hope like hell the thing works."

The Wiffpick team holds collectively a depth of experience that encompasses the history of rocket-astronomy since World War II, from Jim Gunn's nitric acid fireballs in Beeville to the experiments that actually inspired Gunn's Beeville work, namely the V-2 rocket itself. One member of the Wiffpick team, William ("Billy") Baum, is a veteran of Wernher von Braun's V-2 days at the White Sands Proving Grounds. Historians of science credit Baum, along with Richard Tousey and a team of gadgeteers, for putting the first experiment into space. They wanted to capture a spectrum of the sun in ultraviolet light—something that had never been done before, because the earth's atmosphere prevents most ultraviolet light from reaching the ground. They mounted a spectrograph *inside* the explosive warhead in the nose cone of a V-2. ("As to why we left the warhead in the rocket, all I can say is that perhaps we weren't thinking clearly," Baum recalled.) They fired the V-2 into space. The instrument took pictures of the sun until the rocket turned over and fell to earth. The rocket smacked into the desert and the warhead exploded in a tremendous blast and a puff of metallic confetti, and made a crater thirty feet across. For the next try, perhaps thinking more clearly this time, they removed the warhead from a V-2 and mounted a spectrograph on the tailfin of the V-2 in order to improve the instrument's chances for survival, and on October 10, 1946, they fired the rocket. It climbed steadily to an altitude of fifty-five miles while a film snickered through the spectrograph, taking thirty-five good pictures of the sun's spectrum in ultraviolet light. The rocket then yawed out of control and corkscrewed up to an altitude of one hundred miles while shedding pieces of itself. It came tumbling down to earth, and when it hit the ground it made a much smaller crater than the first, containing an intact tailfin from which dangled the world's first space experiment.

Billy Baum, a tall, courtly, soft-voiced, essentially bald man, who is given to wearing cardigan sweaters, is the type specimen of the Palomar gadgeteer. After fooling around with surplus Nazi rockets, Baum came to Palomar Mountain. It was he who mail-ordered the first batch of Army Air Corps electrical hot suits, for a dollar apiece. He built a pulse-counting photometer for the Hale Telescope in 1953—the first sensor ever put on the Hale that could count individual photons, one by one. Baum later went to live on Mars Hill at the Lowell Observatory in Flagstaff, Arizona. When he heard that Westphal and Gunn were planning to build a camera for the Space Telescope, he called them up and told them that he was joining the Wiffpick team. He began attending Wiffpick meetings. In his soft voice, Billy Baum gave the Wiffpickers some advice. He explained that if they put a little black dot on the mirrored pyramid, it would make a black dot in the pictures taken by the Wide Field/Planetary Camera. By moving the Space Telescope ever so slightly until the dot eclipsed a bright star, they might see things lying next to the star that would otherwise be drowned in the star's light—dim planets, brown dwarf stars, rings of comet clouds, solar systems being born. Baum's idea threw the Wiffpick team into a consternation. It horrified some of them. Why, every picture taken by the Wide Field/Planetary Camera would have an ugly little zit on it! That blemish—just think of it. When Wiffpick images were published on the front page of the *New York Times*, people would think that the Wiffpick had a flaw! The Wiffpickers named it the Baum Spot. Billy Baum kept on talking softly about his Spot until the team caved in and agreed to put a black dot on the pyramid—the Baum Spot. "Billy Baum's presence on this team," Gunn said, "is something we have ever been *thankful* for."

▪ ▪ ▪

Shortly after the Wiffpick project got under way, Gunn spent about six months building a CCD camera of his own, an instrument that he named Pfooey. ("Pfooey," Gunn said, "stands for Prime Focus Universal Extragalactic Instrument. PFUEI. *Pfooey*.") Pfooey is a black cylinder containing a single sensor chip, a Nikon lens, and a labyrinth of recycled parts. Pfooey is plugged into prime focus of the Hale Telescope, where Pfooey needs constant tending by an observer who sits in prime focus with Pfooey. Pointed at a bright star, Pfooey can go haywire, provoking the astronomer to yell, "Aw, Pfooey!" In 1979, using Pfooey, Jim Gunn and four other astronomers (Jerry Kristian, Bev Oke, Jim Westphal, and the late Peter Young) mapped the first known gravitational lens. A gravitational lens is a multiple image of a quasar, split and magnified into two, three, or four bright mirages by the presence of a strong gravitational field somewhere in front of the quasar. The gravitational field warps spacetime, fracturing the image of the quasar into blobs. In this case the gravitational field came from a dim, heavy galaxy lying on the line of sight between the quasar and the Milky Way. This discovery is considered to be one of the classics in the annals of modern astronomy.

Megascience dismayed Gunn and Westphal. They joked about kidnapping the Wiffpick. "Gunn and I said to each other, Let's just better move the whole thing down here and just build the goddam thing ourselves," said Westphal. They wondered if they could hide the Space Telescope's camera in an underground maze at Caltech known as the Wastebasket, where perhaps nobody would find the camera until it was too late—after they had knitted the thing together with solder. "Jim and I feel the same way," Westphal said. "Nobody is gonna take my soldering iron away from me."

When Gunn found himself unable to tinker with the Wide Field/Planetary Camera, the *cacoethes gadgetendi* began to gnaw at him. "I had this idea," Gunn said, "that it would be nice to build a camera like the Wiffpick for the Hale Telescope. So I spent a weekend doing a set of preliminary drawings, to see if it would work. I presented the idea to Jim Westphal." They figured costs. The camera looked cheap. They put in a funding request to NASA. NASA agreed to fund the camera out of the Wiffpick budget. Gunn next faced the question of how to get his hands on a tetrad of Space Telescope CCDs—almost a quarter of a million dollars' worth of chips. He hung around the Jet Propulsion Lab, trying to act natural. After what Gunn describes as "a shady chain of events," he managed to get his hands on four chips that had failed their flight qualification tests and would not work in outer space.

Working at night and at odd hours, Gunn calculated formulas for groups of thick mirrors made of slugs of quartz fatter than tuna-fish cans, and hired an optician named Don Loomis to grind them. Gunn also produced a series of engineering drawings for his camera. He lined up Michael Carr, a Caltech engineer, to refine the drawings and to coordinate the building of the instrument. Carr knew a welder who took on scientific jobs, and he persuaded the welder to close up his shop for a day, to roll and weld 4-shooter's steel tube.

Carr felt that an extragalactic camera ought to be painted dead black, and he told Gunn so.

"I want the tube white," Gunn told Carr.

"What about sort of a cream color?"

"Stark white."

Carr eventually got to paint 4-shooter's lid black. Carr built a wheeled cart to hold the tube upright, and he and Gunn pushed the cart and tube into a room next to the Wastebasket.

The Wastebasket is Caltech's answer to the aerospace industry— a warren of rooms beneath Jim Westphal's office on the Caltech campus, inhabited by four engineers. Richard Lucinio, the digital wizard, occupies a central room with his two dogs, who doze on the floor while he taps at a computer terminal. In a corner, under a battery of fluorescent lamps, dwells J. DeVere Smith, once a television repairman and now the wizard of circuitry. Beside a tiny machine shop dwells the wizard of cables, Jovanni Chang. Near the Wastebasket's doorway dwells the wizard of moving parts, Victor Nenow. The wizards maintain passageways to connect their work areas, through which they can walk and receive air, and through which I walked one day, unable to believe the shelves groaning with solder-clotted wires, the cardboard boxes filled with broken electrical units and bits of machinery, the torched and hacksawed circuit boards, the tape recorders, gears, oscilloscopes, keyboards, chunks of foam rubber, magnets, wrenches, telephone books, and toasted and reamed computer terminals. There were dozens of steel cabinets subdivided into small drawers. I pulled open drawer after drawer, and each seemed to hold a different type of bolt or knob or transistor. "If you move something in those drawers"— the wizard of moving parts, Victor Nenow, said in a muffled voice, because he was speaking to me from around a corner—"you are in big trouble, because our short-term memories are no good."

"I really can't stand it down there," Westphal remarked from the safety of his office, one floor above. "I just hope none of that stuff doesn't breed together and start growin'."

In the case of 4-shooter, it did. Gunn walked into the Wastebasket one day and handed around a flyaway stack of papers containing photocopies of pencil drawings (he hated making blueprints; it bored him). He showed the wizards how 4-shooter would be similar to the Wide Field/Planetary Camera—a pyramid would break a shaft of starlight into four parts and reflect the light into an array of CCD cameras. "Gunn doesn't toot his horn," said J. DeVere Smith. "He tells you what he wants to do, and in the back of your mind you are saying, 'That man is going for broke.'" Some of Gunn's sketches of imagined machines inside 4-shooter were barely legible—he had whipped them off during all-night brainstorms. The camera would not have any control knobs. It would contain robots that would take orders only from a computer. Astronomers are clever people, Gunn figured, and would monkey with knobs. He did not want a knob going down some astronomer's windpipe.

The wizards studied Gunn's photocopies. "We have this little problem," DeVere Smith would later say. "We can't read Gunn's writing. It's not a question of doing what he wants; it's a matter of interpreting what he wants." Exercising their option to interpret, the Wizards of the Wastebasket began to build robots that could take orders from a computer. The wizards luxuriated in a certain freedom to improvise. About one third of the machinery inside 4-shooter consists of surplus parts and rehabilitated garbage. It wasn't that Gunn planned to put used parts inside 4-shooter. They just ended up there. "I actually prefer to work with new parts," Gunn said. "But the wizards had their own ideas." The wizards had learned the hard way. They had learned that new parts can fail. For example, they ordered a mini aerospace motor, the size of a roll of quarters, from an aerospace company, at a cost of $125. They plugged the motor into a unit. "It slopped, it grinded, it sounded like hell," Victor Nenow recalled. "We had ordered the wrong kind of motor." It can easily take between eight and thirty weeks to hammer an order for one motor through a corporate marketing bureaucracy, but the wizards knew of a better way. Every lunch hour they pay a visit to a warehouse in Pasadena that sells electrical junk—a place called C&H Sales. "We know the good stuff when it hits," DeVere Smith said. "Sometimes we pick it off the delivery truck before it gets inside the store." At C&H Sales, the wizards found a smooth little Swiss motor in a bin full of motors. They paid five dollars for it, and it ran like a watch. 4-shooter also needed six high-precision stepper motors, some of which would drive wheels that would emplace filters in front of the cameras. Not wanting to take any chances, the wizards asked Gunn if they could go straight to the surplus bins at C&H Sales. Gunn did not mind. At C&H, the wizards found six surplus motors, for which they paid next to nothing, and the motors worked flawlessly.

Most of the common electronic components inside 4-shooter— resistors, transistors, capacitors—came out of bins either in the Wastebasket or at C&H Sales. "Some of the stuff had lived some-where before," Nenow put it. The advanced components for critical circuits—logic chips and gold connectors, for example—Gunn ordered new. 4-shooter needed electricity of different voltages, and so Victor Nenow built a power supply of a type that DeVere Smith describes as "one of Vic's innovations." It is stuffed with fuses, wires, resistors, transformers, and a muffin fan; and Nenow built it largely from nuggets that he dug up around the Wastebasket. Incorporating surplus and junk parts, Nenow built a control sensor that monitors the flow of liquid nitrogen through the cameras. It senses the presence of liquid nitrogen with a ten-cent carbon resis-tor. "We had to order the carbon resistor specially," Nenow said, "because they are so cheap that nobody sells them." J. DeVere

Smith said, "Vic can take *nothing* and make a real piece of scientific gear out of it." As can Smith—he took nothing and built a seismograph from it and installed the seismograph at his home, in order to watch the earthquakes come and go.

The liquid nitrogen tanks on the imaging cameras hang on piano wires that Michael Carr had bought at a warehouse. At one point during 4-shooter's evolution, Gunn installed plastic drive belts inside a spectrograph that would be attached to 4-shooter. The drive belts cracked. Gunn consulted Nenow, who, after some thought, suggested that the belts be replaced with steel-spring drive cables from a movie projector. At C&H Sales, Nenow found some movie projector cables for fifty cents apiece.

Gunn and the wizards believed that American corporations did not want to deal with them. It seemed that nobody was interested in helping Caltech build a supersensitive camera for the Hale Telescope. In Nenow's words, "I had a marketing manager laugh and say, 'We don't want your business.'" Many major American corporations dislike handling small orders from scientists—especially when the scientists are loners, when they have little money to spend, when they are attempting to push a technology to its limits, and who are thus building machines that exceed the comprehension of American corporations. On the customer desirability lists of many companies, gadgeteers from the California Institute of Technology rate somewhere between scrap-iron dealers and the KGB. "Nobody," Nenow said, "wants to send Caltech any trade magazines." Gunn said, "Trying to deal with the front-office people in many companies is almost impossible. Most semiconductor companies won't even send you a catalog. They don't want to send you a catalog. They know perfectly well that you are one guy in a basement at Caltech and that you are going to order one chip. Nobody is interested in selling you one part. They want to sell you a thousand. Then they take four months to deliver. You get beautiful exceptions, once in a while, largely because someone, somewhere in a company, thinks it is a good idea to deal with scientists." Often enough, American companies would accept a small order for parts and never deliver the parts at all. Gunn felt that Japanese firms tended to behave in the same way, at least when dealing with American scientists. Without a trace of irony in his voice he said,

"I gather that things are very much better in Great Britain. British companies seem much more willing to talk with scientists."

While 4-shooter was coming together, Gunn's marriage to Rosemary Wilson deteriorated; another lesson in entropy. He and Rosemary eventually were divorced. Gunn was dating Gillian Knapp. In 1980, she moved to New Jersey to get a job as an astronomer at Princeton University, and Gunn left Caltech to follow her east, where they were married. Jill Knapp and Jim Gunn have become prominent members of the Princeton faculty—they are no longer Calteckers. But a rabbit-warren of rooms in the basement under Jim's office at Princeton has accumulated a lot of electronic junk that has come straight from the Dumpsters of southern California. "At Caltech," Gunn said, "I felt myself turning into an engineer, and that was not where I wanted to go. I also wanted, more than anything else, not to lose Jill Knapp."

Jill Knapp has an interest in the molecular clouds of the Milky Way. She grew up in Dalkeith, Scotland, outside Edinburgh, where her father worked as an industrial chemist. "He gave me a chemistry set," she told me once, "and I am afraid I was one of those wretched kids who are forever making bangs in the kitchen." After she found an old cardboard tube and built a telescope from it, the bangs tapered off. She introduced him to opera. She took him to see a performance of *La Bohème*, and by the last act, the two of them had dissolved into tears. He bought a Sony Walkman, which he carried with him when he traveled to the Hale Telescope. As he gazed at galaxies through the eyepiece of the camera called Pfooey, he listened to *Rigoletto*, *Madame Butterfly*, and *Don Carlo* and ate M&Ms. Soon he discovered Giuseppe Verdi's *Requiem*, a Latin Mass sung for a dead novelist, Alessandro Manzoni. Gunn listened to it over and over, until the tapes hissed and crackled.

> *Sanctus, Sanctus, Sanctus*
> *Dominus, Deus Sabaoth*
> *Pleni sunt coeli et terra gloria tua*
> *Hosanna in excelsis . . .*

> (Holy, Holy, Holy
> Lord, God Almighty

Heaven and earth are filled with your glory
Hosanna in the highest . . .)

He could lose himself in the "Libera Me," at the end of the
Mass, when the voices sang,

Requiem aeternum dona eis, Domine,
et lux perpetua luceat eis.

(Give them eternal rest, Lord,
and may the eternal light shine on them.)

The word *eternal* had no meaning for him—acquainted with the
mathematics of the Grand Unified Theories, Gunn felt that nothing
was eternal, especially temporary things such as space and time.
"The universe," he would say, "cannot survive forever." The universe was the final lesson in entropy. The galaxies would disperse,
while their stars burned down into fogs of black dwarfs and black
holes. After that the basic particles of ordinary matter, the protons,
would almost certainly decay into photons and electrons. The black
holes would also decay. If the universe kept on expanding, it could
conceivably end up as a substance called positronium, some googol
of years hence. Positronium is a very cold plasma of electrons
and antielectrons, bound into loose orbits with one another, each
particle separated by ten million times more space than is contained
in the observable universe today. Gunn had reason to believe that
even the vacuum itself might be unstable, liable to catastrophic
decay. Thus, like a bubble, the universe might one day pop and
vanish. Not even *nothing* could last forever. The universe might
not even be unique. Why should there be one universe? Nature
never stopped at one of anything! Not when an octillion models
might work. This universe might be a bubble drifting on the Planck
soup, one of an infinity of other universes that were constantly
erupting out of the Planck soup, by random chance. Maybe God
plays at dice with universes. But when he was cocooned in prime
focus with Giuseppe Verdi and a value-pack of M&Ms, staring
into the mirror, where he could see with his own eye a warp in
spacetime in the form of a gravitational lens shining like a pair of

headlights out of the past while the Big Eye tipped westward through the tremendous unvalving doors of night, Gunn felt that there was a kind of salvation in light.

Jill Knapp worried about Jim up in prime focus. She worried that he might die in prime focus, a victim of hypothermia, sleeplessness, and chocolate shock. She believed that there were worse ways to go.

After he had settled on the East Coast, Gunn flew back to Caltech every few weeks to build a unit or to solder a circuit board himself, while gradually 4-shooter's white cylinder accreted parts. The Hale Telescope had always been guided in the traditional manner: the astronomer held a paddle while watching a set of crosshairs and a guide star. Gunn thought that a robot could do better than the human hand, so he built a robot to guide the Hale Telescope and tucked it inside 4-shooter. An arm holding a pick-off mirror—it resembles a mirror on a dental tool—reaches into the shaft of starlight coming into 4-shooter and picks off the gleam of a single star: the guide star. The mirror shines the guide star's light through a miniature telescope—a telescope big enough to hold one star. The telescope focuses the starlight onto a spinning razor blade. The razor blade chops up the light, making the star seem to blink. A small computer, which Gunn built by hand, watches the blinking star. The computer interrogates the guide star. It asks: Is this star drifting? Where?—and orders the Hale Telescope to correct its motion accordingly. The razor blade is a Wilkinson Sword that Gunn bought at a Rexall drugstore in Pasadena. He broke the blade to the right shape with a pair of pliers and fastened it in place with a dab of glue.

Most of the electrical wiring inside 4-shooter is Teflon-coated wire, a type often used in aerospace applications, because it does not easily melt. It is expensive stuff. The wizards bought their Teflon wire at C&H Sales for a half a cent a foot. Since 4-shooter contains something like one mile of wiring, the savings added up. Victor Nenow saved a bit more by stripping Teflon wires out of dead computers. When Nenow had nothing better to do, he broke open dead computers. While he hacksawed a circuit board and nipped wires out of it, he toasted cheese sandwiches in a toaster oven that he had built for himself—another of Vic's innovations.

He showed me how it worked. He slid a cheese sandwich onto a heat sink, under a grill. He turned a dial. The oven glowed. He cranked up the dial. A blinding light flooded from the toaster oven—the heating elements were four movie studio lamps. He put on a pair of sunglasses and said, "That's only half voltage." He had marked the dial to avoid damaging his eyes or igniting the cheese. "I used to electrocute hot dogs," he said. "Put a clip on either end of the hot dog and send a current through it. The hot dog would steam and split. It worked great."

The administrators of the California Institute of Technology have not the slightest official idea of how to classify the Wizards of the Wastebasket.

"Our job titles?" DeVere Smith wondered. "Vic!" he called out, "what are our job titles?"

Nenow's answer filtered around a corner: "We're garbologists."

J. DeVere Smith, who etched and soldered many of 4-shooter's circuit boards, is a tall, white-haired man with large, deft hands. In 1930, Smith opened a radio repair shop in Los Angeles that he deemed Advance Radio, although he admits that he did fix the occasional Victrola. Advance Radio also handled meat—delivered it to grocery stores (a profitable sideline). Smith eventually got out of meat and radios and moved into television repair, although he continued to call his shop Advance Radio. He finally sold his shop, and as he puts it, "I retired and came up to Caltech. When they get through with me here, they'll have to bury me." Shortly after Smith arrived at Caltech, he and Victor Nenow utilized the science of garbology to build the electronic systems for four mass spectrometers that Caltech geophysicists used to analyze moon rocks gathered during the Apollo lunar landings.

An astronomer walked into the Wastebasket and threaded his way to DeVere Smith's corner. "Hi, DeVere," he said. "I need a knob."

"You've come to the right place," DeVere said. He pushed some things around his workbench until he had found a knob. He said, "How's that?"

The astronomer inspected it. "De*Vere*. This is trashy. I want a shiny knob."

J. DeVere Smith slid open a legal-sized filing drawer by his knee.

The drawer was absolutely packed with knobs. "You want two?" he asked.

Smith is a miner of dumpsters. "You'd be amazed," he said, "at what you can find in dumpsters." I was amazed. One day I was hanging around the Wastebasket, talking with the wizards, when a geologist walked in.

Nenow held up a bag and pulled a ceramic cylinder out of the bag. He handed the cylinder to the geologist. He said, "I thought maybe you could tell me what this is."

The geologist turned the object over. "Oh! It's a proton precession magnetometer."

"A what?"

"It measures the strength of the earth's magnetic field. This is a pretty good sensor. Where did you get this?"

"It wound up in a trash bin. DeVere found it."

"You mean somebody threw it out?"

"Oh, sure."

"Can I keep it?"

"Sure."

"Thanks. These things are fifty thousand dollars each."

Exit geologist, carrying sensor.

Richard Lucinio, the digital wizard, preferred to work at night. He would leave his home in Topanga Canyon in late afternoon, with his two dogs, and drive to Caltech. He designed the logic boards inside 4-shooter. These circuits contain chips used to control both machines and other chips. Lucinio's logic boards, for example, can order the CCDs to dump their electrons into the amplifiers. Lucinio would collect a robot from one of the other wizards as the wizard was leaving work. He would hook the robot into a logic board and play with it all night, trying to get a motor to start or some wheels to turn. He often worked with Barbara Zimmerman, who wrote the software program that controls 4-shooter. When they could not get a robot to work, they would have a glorious time of finger pointing: "It's the hardware!" "Naw, it's the software!" They would try various command sequences on the robot, cajoling it. Eventually the robot would wake up. Jovanni Chang (who soldered and tied together many of the cable harnesses inside 4-shooter) liked to watch these miracles. As he described it, "We would hear a *whee! Chunk!* Something would happen—a motor would spin, a trapdoor would open. It was like a flower blooming."

Gunn built a set of amplifiers, to process electrons coming from the CCD chips. Some of the circuits he soldered himself, and some he gave to DeVere Smith to solder. On a Saturday afternoon in September 1983, Gunn and Michael Carr rented a Ryder truck and stuffed 4-shooter into it. Carr took the wheel. Four years of Carr's life were in the back of that truck, and he drove the Ryder at thirty-eight miles an hour down Interstate 210. Gunn asked to drive. Carr exchanged places with Gunn, which he immediately realized had been a mistake, because Gunn put the Ryder in the fast lane and mushed the accelerator. ("Gunn was overanxious," Carr thinks.) When the Ryder headed up the switchbacks of Palomar Mountain, Michael Carr experienced a touch of pure fear. One hour after dusk they had loaded 4-shooter into the Hale Telescope, and 4-shooter's chips had begun collecting photons. A nearby dwarf star contributed first light to 4-shooter, and then Gunn aimed it at galaxies, running for the deep range. "He had that little Gunn grin on his face," Carr remembered. "I just hope I never lose sight of Jim Gunn."

Not long after its installation in the Hale Telescope, 4-shooter entered unexplored spacetime. The camera exposed images of galaxies that were deeper in the universe than any galaxies that had ever been seen before. These galaxies were different from nearby galaxies. They seemed to be bluer in color; saturated with hot, young stars. 4-shooter was seeing galaxies in an earlier cosmic epoch. When 4-shooter was installed in the Hale Telescope, the Hale became at least one hundred times more powerful than it was when it was first built. In order to equal the light-gathering power of 4-shooter coupled to a two-hundred-inch mirror, George Ellery Hale would have had to build a telescope with a mirror at least one hundred and sixty-six feet across—a mirror the size of a parking lot.

The news of what Gunn was doing underneath Caltech must have filtered into the scientific community, because shortly before he finished 4-shooter, Gunn received a telephone call from the John D. and Catherine T. MacArthur Foundation, informing him that he had won a MacArthur Fellowship—one of the so-called genius awards. Gunn would collect a slug of money totaling $220,000 over the next five years, which he could spend any way he pleased. He immediately telephoned his wife, who happened to be on a run at a radio telescope.

"Oh, Jim!" she said. "We can get a box at the Met!"

"This is serious."

"That's what I'm talking about," Jill said.

A box at the Metropolitan Opera seemed a bit extravagant for a couple of astronomers, but they did buy season tickets. Apart from that, Gunn could not figure out how to spend the money. He bought a laser-disk video player in order to watch opera at home, and he wired it into his homemade stereo. He gave some money to graduate students to help them with their travel. The rest of the money is sitting in a bank account. He might have invested some of it in a good pair of prescription reading glasses, but, in his words, "They are tremendously expensive." He enlarged his Woolworth collection.

▪ ▪ ▪

Maarten Schmidt had been watching Gunn from afar. Although trained in classical photography, Schmidt was not unaware of this person who could build an extragalactic camera from out-of-the-Dumpster parts. By the time Gunn had finished building Pfooey, Schmidt wanted to get a collaboration going with Gunn, but then so did a lot of other people. Schmidt had his own reasons.

In 1967, Schmidt had discovered that as one looks out into space the universe appears to become thickly settled with quasars. He showed that quasars were a pathological aspect of the young universe and are now largely extinct. He wanted to collect a small but carefully selected sample of quasars, in order to understand the characteristics of quasars as a population, in much the way that one would sample the opinions of a few people in order to understand the mood of a nation. Working with a colleague named Richard Green, Schmidt directed a long survey for quasars, using the eighteen-inch Schmidt telescope on Palomar Mountain. Green eventually found ninety-four quasars. When he and Schmidt analyzed their data, they discovered that the population of brightest quasars had burned out sharply and abruptly as the universe evolved, while the less brilliant quasars had malingered. But the Little Eye could not pick up quasars with deep redshifts. Schmidt wondered just how many deeply redshifted quasars might be hidden among the stars.

Schmidt and other quasar hunters had noticed that quasars seemed difficult to find at great distances, at enormous lookback times. They sensed a falling off. As they explored deeper, they found fewer and fewer quasars, and the redshifts trailed off: 2.69 ... 2.75 ... 2.88 ... 3.40 ... 3.78. In 1987, a team found a quasar with a redshift of 4.43, which corresponded to a rough distance (only guessed) of thirteen billion light-years. "You must have perceived a slowing down," Maarten said to me on the catwalk one night. "Why don't we see quasars with redshifts of five, six, or seven?" Watching astronomers trying to discover the most distant quasar was like watching javelin throwers trying to break a world record; the javelins were landing close together, approaching a natural limit. Astronomers began to feel that they were looking through a veil of quasars into darkness—into the edge of the universe. The redshift cutoff.

An astronomer named Patrick Osmer found the first scientifically clear evidence that, as he looked past a redshift of 3.5, he had seen the redshift cutoff begin. But the fine structure of the redshift cutoff remained an enigma. Some astronomers suspected that only the Hubble Space Telescope would be powerful enough to probe to the edge of the universe and thereby to define how and when quasars caught fire. But to reach the redshift cutoff with the Hale Telescope would, for Maarten Schmidt, conclude an odyssey into lookback time that had begun accidentally one afternoon in 1963, when he had dusted a little glass slide with a handkerchief and put it into a microscope.

In 1982, Schmidt asked Gunn rather casually if Gunn might be interested in using Pfooey to map the outer structure of the universe. Gunn thought that sounded like a nice idea. Schmidt immediately pulled Don Schneider into the experiment; Don was then Schmidt's assistant. Schmidt, Schneider, and Gunn began their search by taking more than one hundred electronic snapshots through a prism installed in Pfooey. These snapshots revealed some quasars, but none were particularly near the edge of the universe. Schmidt became restless. The Pfooey snapshots did not cover much sky. He began to think about dramatic solutions. One night when they were taking snapshots, Maarten asked Jim if Pfooey could be turned into a scanner and swept across the sky.

"That was obviously a big job," Maarten would recall, "because Jim had to think about my question for five minutes."

Jim put his hands over his face. He ran Pfooey's control program through his memory. He took his hands from his face and said, "Yes, Maarten, we can do it."

Merely by changing a few lines of computer code, Gunn turned Pfooey into a scanner, which worked fairly well, except that every time a bright star passed across Pfooey's sensor chip, it sent electrons caroming like golf balls through Pfooey, which was a pain. The astronomers found a few quasars, but the highest redshift came to only 2.76. A decade ago that would have been a satisfying redshift, but today it was a disappointment. They felt sure that their technique had been good—they had not found any remote quasars simply because such quasars were extremely rare. This surprised them. They concluded, tentatively, that the redshift cutoff was nearby and abrupt; and that therefore the population of quasars had been born suddenly and rather late in the development of the universe. But Schmidt had his doubts about all this. He wondered if there might be a hidden population of quasars buried deep within lookback time, near the beginning of all things. When Gunn finished building 4-shooter, Schmidt asked Gunn if he would not mind turning 4-shooter into a scanner, since with a quartet of cameras it could devour sky. Schmidt felt like a paleontologist walking through a dessicated African valley, sensing the presence of fossil beds under his feet that might hold teeth and skull fragments of missing links and half-imagined creatures; but whether luck and his tools and his hopes would serve him well or not at all, he could not say or know. The edge of the universe haunted Maarten Schmidt, and he now felt that he no longer understood the redshift cutoff at all.

PART 4

Discoveries

Richard Lucinio liked to find reasons to drop by Palomar Mountain. He showed up on Palomar at two o'clock one morning, obviously traveling on business, because he was carrying an attaché case. He stood in front of a video screen, watching a river of galaxies slide past, and said, "Gunn, I have never seen anything like this. Those things are like grains of sand out there."

"There are lots of areas of the universe that look like this," Gunn said casually. "Places where the sky is covered with galaxies. But just because you see beaucoup galaxies in one place and only a few galaxies in the next place, it doesn't mean that that's a large perturbation in the structure of the universe."

After reflecting on this, the digital wizard went into a small workshop next to the data room, carrying the attaché case. His two dogs followed him. He opened the attaché case, and there was a computer inside it—keyboard, chips, illuminated display. A smell of burning solder drifted through the data room. "A wizard's work," Lucinio said, "is never done." He said that the attaché case was an emergency control system for 4-shooter. "I can hook this thing into 4-shooter and run all the robotics with it," he said. His dogs grew bored and wandered around the data room. One dog fell asleep underneath the table that held the main video screen. He had a dream, and his paw pads wobbled. He thought he was running.

One evening Jim and Maarten got into a discussion about television. Most of the TV programs that Jim watched were videos from the Hale Telescope. Maarten, on the other hand, enjoyed late-night television—it helped to clear the quasars from his mind. "I

don't imagine you saw the World Wrestling Federation tourna-
ments, James?" he wondered.

"What?" Gunn said.

"They were last week. On late-night television."

"I am afraid I missed that, Maarten."

"That's too bad. There was this free-for-all. It was involving these
exceedingly terrible guys, this motorcycle gang. They seemed to
be hitting everybody with chains, including the referees. It was
just incredible. Then you should have seen what happened to—
what do they call this guy? Ja! Hulk Hogan! This other guy—I
mean, this really enormous fat guy—got mad at Hulk Hogan. You
could see him getting madder and madder, and then he ran across
the ring and just *fell* on top of this Hogan, rendering him unconscious, as it were. After that they showed him being loaded into
an ambulance—"

"Why do you watch that stuff, Maarten?" Jim asked.

"It helps me get to sleep."

"I don't see how."

"Part of the problem is that the television seems exceedingly
difficult to turn off."

"There's a switch on it, Maarten."

"That's too much trouble."

"There's also a plug on the wall."

"Ja, but who wants to get out of bed for that?"

To pass the time while they watched galaxies, Gunn and Schneider held science-fiction quizzes.

" 'The planets you may one day possess,' Don quoted, 'but the
stars are not for Man.' Who said that?"

" 'The stars are not for Man.' I know that line," Gunn said. He
walked back and forth. "Karellen. Karellen said that, in *Childhood's
End*. By Arthur C. Clarke. One of the best science-fiction books
ever written."

"Yes. It had to be Karellen," Don said, "because he was the only
one who talked of deep motives—and he was the alien. But *Dune*
is the best science-fiction book ever written."

"No, no, Don, *Childhood's End*."

"I agree with you that *Childhood's End* is a great book," Don
said. "But 'The stars are not for Man'—those are the saddest words

I have ever read. And do you remember these words: 'Yes, we have had our failures'?"

Gunn paced the room with his hands in his pockets. He was wearing his standard eyeglasses, the pair with electrical tape around the nosepiece. His down jacket piled up in a sort of hunch over his back. "Sure!" he said. "That was also Karellen!"

Juan Carrasco, who generally listened to such discussions without comment, smiled and said, " 'We have had our failures.' I like that." The astronomers grinned at the night assistant and said nothing. Yes, Juan had seen them all.

I asked Gunn one night, "Do you think there is life in these galaxies—somebody out there scanning the Milky Way?"

"Well . . ." A gleam entered his eye. Glancing at Don Schneider, Gunn said, "Maybe not in *every* galaxy. Maybe somebody in every *third* galaxy is looking at us."

Don Schneider turned a level look on Jim Gunn. Don was shocked, or professed to be, because he didn't believe in the existence of intelligent life out there. With a honed edge in his voice Don said, "It's amazing, Jim, to think that we really are alone in this big universe."

It was Gunn's turn to be shocked. "Why, Donz! Don't you believe it for a moment!"

"I am a scientist. I believe evidence. I will believe any *evidence* you care to show me regarding the existence of aliens."

"Evidence! All the evidence to show there is *no* intelligent life out there amounts to absolutely *nothing*."

"I have locked onto a debating shyster," Don said.

"It is amazing, Don, to think how crowded this universe is."

"I am ready to be convinced."

"You will."

"When?" Don demanded.

Jim Gunn did not answer. He smiled archly and sipped a Von's Lemon-Lime soda.

Don turned to the night assistant for support. "You see, Juan, as with all my debating activities, my opponents just turn their backs, crushed in utter defeat, yet affording me no satisfaction whatever."

"Yes," the night assistant agreed, "that is depressing."

Gunn said, "The number of civilizations out there is—"

"Zero," Don interrupted.

"Infinity."

"Well, the answer is one of the two," Don said, abruptly sounding less certain of himself. He said, "Have you seen that *Twilight Zone*, 'To Serve Man'?"

"Terrific story," Jim said. "I read it."

"Well, then."

"Well, what, Donz?"

"Well, as you know, the aliens came to earth, and they brought a book with them. Do you remember the title of that book, Jim? *To Serve Man*. Everybody was so happy about that. Then doubtless you remember what happened." Don gave a rather involved plot summary, the punch line being that *To Serve Man* had been a cookbook.

Gunn did not reply.

"Also," Don asked, "where is everybody?"

Gunn nodded toward the screen. "Right there."

Don said, "You have to admit that my point of view is equally defensible."

Don Schneider was a devout Roman Catholic, and he preferred to think of the human species as a chosen people, and space as the final frontier. He believed that humanity was slowly readying itself for what many people involved in astronomy felt was the inevitable step. They called it the Breakout—that we would reach for the stars, not with a mirror but a ship. "Space," he said, "is our manifest destiny." What bothered him was the thought that if alien civilizations existed, then some of them would have advanced a few billion years ahead of us. One billion years amounted to four rotations of a galaxy, four galactic years. *Homo sapiens* had lasted, so far, for just forty minutes out of a galactic year, whereas a culture two billion years ahead of us would be eight galactic years old. "Powerful cultures," he said, "always destroy less technologically advanced cultures. When the Europeans landed in North America, they were only a few thousand years ahead of the Indians, and look what happened to the Indians." Don believed that a contact with aliens might be the greatest danger we would ever face. A global nuclear war or a plague from a hot virus would kill a lot of people but would not be likely to kill the human spirit. He felt that even to believe in the existence of alien civilizations would be to erase the purpose of his life as a scientist. A two-billion-year-old culture would be four times as distant from us in time as we are from the trilobites. If there was somebody out there, what would they care for the ambition of a slime mold?

Gunn, for his part, wondered what he would do if a civilization offered him two billion years' worth of science. He imagined that knowledge as a symbolic language written in a book—a book with all the answers in it. "Would I open that book?" he wondered out loud. "I don't think I would open it. Well, I would not be able to resist the temptation. We would read it, but we would not understand it, although we would know it was written by somebody who did understand, and that would kill us."

Maarten remained silent during the alien debates. He preferred to stick to conservative questions that he thought he could answer with a telescope, such as that of the unpublished history of the universe.

▪ ▪ ▪

One night the astronomers were watching spectra float up the screen, the way one likes to stand on a bridge over a lazy stream and observe different shapes and species of leaves drift by on the water, while Juan Carrasco was watching the temperature of the mirror. He said abruptly, "The humidity is going up." He was worried that a dew might fall on the mirror.

The astronomers came over to Juan's instruments.

Maarten said, "We are barely over the dewpoint." A warning light began to blink. Maarten said, "Those numbers are falling like crazy." A buzzer went off. "We have to close," Maarten said.

Juan threw a switch. "Mirror closed."

The galaxies snuffed out.

Juan climbed to the catwalk. He saw a narrow moon. He extended his arm and placed his thumb over the moon. A glory of faint cirrus clouds haloed his thumb. He slapped the outside wall of the dome. It felt wet and cold. He returned to the data room and announced that the sky held structure. The astronomers talked and decided that the clouds would last. In that case, Juan said, the astronomers could find him downstairs. Pulling a ring-binder notebook from a shelf, he said, "I am going to do motor inventory."

The Palomar engineers were attempting to "characterize" the Hale Telescope, which is another way of saying that they were trying to figure out how it worked. Juan had been given the assignment of locating every motor in the dome. The ring-binder notebook contained a long list of motors. "It is my job," he explained to me, "to determine which of these motors exist." He pulled on a pair of rubber boots and took the elevator down one level. I followed him. We crossed a walkway and arrived at a set of electric pumps that drove oil up into the telescope.

We were deep in the bowels of the dome. He turned on his flashlight and pointed it around, into dim areas, across cabinets covered with dials and racks full of radio tubes. This was the graveyard of instruments. When a gadgeteer's machine had outlived its usefulness, the astronomers left it there to collect dust and possibly to serve as a source of parts. Someday 4-shooter would be sitting here. Juan consulted his notebook. He stepped into a drip pan that contained puddles of Flying Horse telescope oil. He said that during motor inventory, he might find that a motor listed in the blueprints had never been built. Or he might discover a mystery motor that had been running calmly since Harry Truman was president. He bent over and read a serial number on a motor. "That checks out," he said. He walked around, poking at things. He said, "I have found five Vickers pumps here, but there are supposed to be six. I can't locate the sixth pump. It may not be here at all. There is some confusion about what exists in this telescope and what does not exist." He touched the brim of his hard hat and pointed his flashlight around.

After a while I asked, "How well do you know your way around the Big Eye?"

"I *lived* with that telescope."

"Can you get inside it?"

He smiled. "Have you never been inside the Big Eye?"

"No."

"This is enough motor inventory for tonight," he said.

He led me upstairs. We stopped at the base of the telescope, looking up. The Hale Telescope's tube hangs between the prongs of a fork, called the yoke. The arms of the yoke are called the East Arm and the West Arm. Each arm is ten feet in diameter. Juan stepped through a circular doorway into the base of the West Arm. I followed him. He flipped a light switch. He pointed to a small porthole at his feet. "One can climb down in there," he said, meaning that I ought to. I lowered myself through the porthole, and then climbed through three rooms, separated by bulkheads. There were pools of oil in the rooms—Mobil Flying Horse telescope oil, number 95. I rubbed some on my fingertips. It was clear and golden in color, and had a sweet smell. Juan said, "This oil fills up with moths in the summer."

I climbed back out, and then Juan led the way up a stair that angled upward for three stories through the West Arm of the Hale Telescope, until we came to a landing with a cabinet standing on it. The cabinet contained a set of machinery that controlled the telescope's motion north and south—the motion through declination, or latitude. He lifted a cowling from the cabinet, exposing clumps of motors, gears, and boxes leaking oil. This was a mechanical computer. (I learned afterward that a young engineer named Sinclair Smith had begun to design it in the 1930s. Smith had died of cancer. Bruce Rule, who designed the mirror-support machines, had finished Smith's job.) Juan said that the observatory had recently installed digital computers to take over the work of these mechanical computers, but the observatory kept Rule's computers oiled and ready for use, in case the digital computers crashed.

Juan replaced the cowling over the computer and said, "This is like a submarine." His voice echoed down the West Arm.

"Or like Buck Rogers's spaceship."

"It is literally a starship," he said. He squeezed behind the mechanical computers and suddenly pulled himself upward and disappeared into a porthole.

I followed him through the porthole, into a greasy passageway. We climbed upward through the telescope, and finally emerged into the upper half of the tubular West Arm, where there were no more stairs. Juan pointed to a hydraulic assembly. "That's a clutch," he said. "It slips sometimes. Then I have to climb up in here and fix it in a *big* hurry."

"Because the astronomer is screaming?"

"They usually don't scream," he said. "But they are ready to do so."

We crawled upward on hands and feet, which produced faint, hollow booms in the arm.

Having watched the Hale Telescope evolve during the twenty years he had served it as caretaker, Juan sensed its personality. He did think it had a personality: good-natured but quirky. On one occasion, every time he had pushed the "north guide" button, the telescope had moved *south*. Looking around the chamber we were in, he said, "There is one loose bolt in here somewhere. When you slew this telescope way over to the west, you can hear that bolt slide from one end of the telescope to the other." Also, a door would sometimes slam unexpectedly inside the telescope, which would shake the stars. And the astronomer would be ready to scream.

At the top of the West Arm we arrived at a bulkhead containing a porthole. I put my hands around the lip of the porthole. When I looked back over my shoulder, I realized that we had climbed a long way up the West Arm.

"I am not afraid of heights," Juan said, gripping the porthole. "But I respect them."

I put my head through the porthole and looked around. A view opened downward into a warren of dark chambers. Juan said that we were looking down inside the horseshoe bearing, which was hollow. He had climbed around in it, he said, but, "I don't think you would want to."

The Hale's horseshoe bearing is the largest bearing ever made— a C-shaped arc forty-six feet in diameter. It contains nearly half a mile of welded seams. The load-bearing outer curve of the arc floats on a film of Mobil Flying Horse telescope oil. The moving parts of the Hale Telescope weigh 1.1 million pounds and are as exquisitely balanced as the escapement of a watch. Bob Thicksten once disengaged the telescope's clutches and counterweights and stood in the West Arm to see what would happen. The telescope began to tip westward, ever so slightly unbalanced by the weight of Thicksten's body. Thicksten believed that if he had stood in the West Arm long enough, the telescope finally would have turned over sideways. The motor that drives the entire telescope in tandem

THE · PRIME · FOCVS ·

TWO · HVNDRED · INCH · TELESCOPE

A view of the prime focus cage at the Hale Telescope, drawn by Russell W. Porter. We are looking down the tube of the telescope toward the main mirror. An astronomer is sitting in the prime focus cage staring into the mirror. At Palomar you sometimes hear rumors of the remarkable Prime Focus Club, a mysterious and exclusive group of astronomers who claim to have made love with someone while in the prime focus cage of the Hale Telescope. This kind of thing would most likely happen during a cloudy night, when the astronomers are bored and have nothing to do. (Photograph courtesy of Palomar/Caltech) with the moving sky is a one-twelfth-horsepower Bodine electric motor the size of a grapefruit, made in the U.S.A. around the year 1942, and it has never been replaced.

Juan began to wonder what the weather was doing. We crab-walked down the West Arm, descended the stairs, and climbed out of the telescope onto the floor. Jim Gunn appeared.

"Juan, we need you," he said. We hurried into the data room. The sky had cleared. Forbidden to touch the controls, the astronomers had been pacing around the data room, wondering where Juan had gotten off to.

Juan hit a bank of toggles. "Mirror opened," he said. "We are looking out."

▪ ▪ ▪

When Juan Carrasco was a boy and looked out, he believed that the sky was a hard bowl over the earth, that on the other side of the bowl lay heaven. He spent his childhood in a one-room adobe hut in Balmorhea, Texas, with six brothers and sisters, and like them, he was baptized by the parish priest, Father Salvador Girán. Juan's father, Apolonio Carrasco, had built the hut with his own hands. When Juan was nine years old, Apolonio managed to get a loan from the Farmers' Home Administration to buy a farm outside Balmorhea: sixty acres of black dirt and a small wooden Anglo house. In the cold winters of west Texas, the family burned mesquite in a stove in the living room, where they took their baths in a galvanized tub. The main piece of furniture in the living room was a bed, which was actually a couch for visitors. The Carrascos owned six cows, one lamb, and a few chickens. In a good year Apolonio raised twenty bales of cotton. In a bad year, "Bueno, hicimos el vivir," Apolonio would say with a little shrug and a smile—"Oh, well, we made a living."

Carrasco is a scrub oak with a tough silver-green leaf, serrated with spines. It grows in clumps of chaparral on the sun-facing slopes of mountains throughout the southwest, and it tends to prosper in places where other trees do not. "Those Carrascos, they are rich," went the talk around Balmorhea. Juan was not so sure. Other kids skipped school to earn a dollar a day picking cotton. Juan's parents made their children stay in school. "Those other kids can buy a pair of Levi's and maybe a new shirt," Juan said to himself. Balmorhea had one gas station and a movie theater, "but we hardly had enough money to go to the theater on Saturday night." On big nights Juan and his brothers dressed up and walked over to a place called The Country Club. "It was not the kind of country club you are thinking of. It was a cantina. We danced and drank beer." Apolonio avoided The Country Club, thereby earning himself a reputation for being too serious about himself and his farm. When he went into Pecos to buy equipment or to borrow money, he put on a ten-gallon Stetson hat and gabardine trousers, and he tucked a silver-plated Colt .38—a gun known all over Pecos—at the small of his back. That gun had killed plenty of men. It had once belonged to the City Marshal of Pecos. Apolonio never tried to shoot anybody with it; sometimes he would blast it at a coyote hanging around the chicken coop, but could not seem to hit the coyote with it. His gun was a piece of jewelry; he wore it to weddings. All the men wore their guns to weddings. They locked the guns in their trucks before they went into the church, out of respect for Father Girán and the Holy Spirit, and later, at the wedding feast, they fired their guns into the air—*vivas* to the good luck of the newlyweds.

Apolonio planted watermelons among his rows of cotton. The crows liked the watermelons. They would walk around and poke at them. Juan's mother, Ysabel Carrasco, would cry out to Juan's oldest sister, "Aurora, Aurora! Traite el quate"—"Aurora, get the twin."

Aurora would come running with a shotgun (the "twin") and fire a load of buckshot over the heads of the crows. The kick would knock Aurora back a few steps, and the crows would take off with sullen, disconsolate flaps. It was necessary to guard the watermelons all the time, because they made cotton picking bearable. Juan remembered how sweet they were. "You could hit one of the little ones with a knife, take the heart, and wipe your face on your sleeve." At the end of a day of cotton picking, Ysabel would line up her sons and peel off their watermelony shirts and drop the shirts into a gasoline-driven Maytag washer that stood in the backyard. "Aurora!" she would cry. "Commence the machine." Aurora, who was nothing if not a gadgeteer, would pull out the Maytag's spark plug and put a drop of gasoline in the engine's one cylinder, put the spark plug back in, give the Maytag a terrific kick-start with her foot, and the Maytag would "commence" with a roar. It was deafeningly loud, and it blew flames out through its exhaust pipe. It washed their clothes in a hurry.

Summer nights were too hot for anybody to sleep. The whole town of Balmorhea would be awake. The Carrascos sat on their front porch until the early hours of the morning, hoping always for a visit from the parish priest, Father Salvador Girán. Whenever Father Girán dropped by, the Carrascos put him in a chair of honor on the porch. Then Juan scrambled to put his chair next to Father Girán.

"Would you like a glass of water, Padre?"

"Gracias, Juanito."

The priest was of modest height but strong, from working with adobe, and he was a Spaniard by birth. He had milky Spanish skin and gold-rimmed spectacles, and he wore a black suit with a Roman collar. He had received a doctorate in physics in Spain. Then, for some reason, he had left physics to become a priest, and had gone to South America as a missionary, and then he had come to Texas. He told stories far into the night—about his travels in Spain and South America, about how to find water with a prayer book and a skeleton key; about the latest progress on his new adobe church on the grounds of the Mission of Our Lady of Guadalupe, which he was building with his own hands from adobe bricks; about an old Spanish lady who had once tried to poison him with a hard-boiled egg. Father Girán knew the constellations. "The Cross is high," he would say late on an August night, and the Carrascos would lean off their porch to look.

During a lull in the conversation someone might notice a shooting star. The Carrascos asked Father Girán where a star went when it fell off the sky.

Stars don't go anywhere, he told them. If a star hit the earth, there would not be anything left of the earth.

Then did they drown in the sea?

No, no! He laughed. They do not drown in the sea. The stars, he said, are enormous and far away.

This surprised the Carrascos, especially Juan. But if Father Girán said this was true, who were they to question him?

"The stars are in outer space," he said. "They are a great distance from the earth."

Juan tried to imagine a space with stars floating in it. He said to Father Girán, "I wish I knew what a star looked like nearby."

"You already do," Father Girán said. "The sun is a star."

The response of the Carrascos was dead silence. It was a silence of disbelief.

He said that the sun, as a star, was an extremely hot ball of gas, and that its heat brought life to the earth; that, he said, was why the land produced cotton. The sun and the stars, he said, were many, many times bigger than the earth.

The sun is bigger than the earth? Juan wondered. The stars are bigger than the earth? Then where does a star go, Juan asked, when it falls?

The stars never fall, he said to Juan. You are thinking of a meteor—a *pajita*. He used the word *pajita*, which means "little straw," because a meteor looks like a straw blowing in the wind. He explained that a *pajita* is a very, very tiny pebble coming in from outer space. *Pajitas* burn up through friction in the atmosphere.

Apolonio and his sons had seen *pajitas* that lit up the whole sky when they were irrigating the fields at night. They mentioned this fact to Father Girán.

"One of those really bright *pajitas* might be the size of a marble," Father Girán said.

That amazed the Carrascos.

Father Girán said, "Juanito, if you are with your father in the fields and you see a spectacular *pajita*, you should be very quiet and listen. Try to hear something drop. You might even find one on the ground."

Juan often worked with his father in the cotton fields at night, when they would irrigate the crops. Sometimes Juan would point to a star and say to his father, "Can you believe that that star is many times larger than the earth?"

"That I would never have known," Apolonio would say. Then Apolonio might say, "Mire! Es una pajita!" and he and Juan would freeze, listening. "I wish the water weren't so noisy around here," Apolonio would say. But even when the water ran quietly in the ditches, they never heard the sound of a shooting star, and they never found one lying in the fields.

Juan dropped out of school in ninth grade because he wanted to get a job. The job he found was that of working in the fields with his father. When the Korean War began, he was drafted into the Army and sent to guard the city of Pittsburgh. By the time he returned to Balmorhea, he wanted to get a salaried job, and so he went to San Antonio, where he enrolled in the Lewis Barber College, which was then located in a condemned building at 124 East Military Plaza, an easy walk from skid row. He started his training in the back of a vast room full of chairs, giving free haircuts to winos.

The San Antonio winos were the slickest-looking bums in Texas. They populated the back chairs at Lewis Barber College. Juan and the other beginners began the treatment on a wino by giving the wino a shampoo. When you tilted the guy back in the chair, you could often hear a bottle of muscatel gurgle in his pocket. After the shampoo Juan would cut his hair. He would shave him with a straight razor. He would give the wino a mud pack on the face. He would rinse the mud off the wino's face, and then he would vibrate the wino's cheeks and neck with an electric machine that broke up fat deposits under the skin and enhanced the complexion. He would oil the wino's hair with Wildroot Cream and give him a splash of rum aftershave. Then the wino would look in the mirror and say, "Damn! I didn't know I was such a good-looking guy."

The college's head professor of razors was a ferocious, seized-up old Texan named Patterson, who had either a wooden leg or a bad live leg—nobody was sure which and nobody dared ask him, since they said he had been in the Texas Rangers before he had wrecked his leg, and he carried a razor and knew how to use it. Patterson would limp around the back of the room, on the prowl for a student botching a job on a wino. "You be careful with those guys," he said to Juan. "You keep the rum lotion locked up, you hear? Now, your customer might want to cough or jerk around or something while you're shaving him. You could cut him real bad. You could infect his face." He watched Juan prepare a wino for shaving. "What do you think you are doing?" he screamed at Juan. "Get that towel off his face! You're smothering the customer!"

Patterson opened his razor and fixed an eye on the wino, who fixed an eye on Patterson's razor. Patterson stooped over and took up a leather belt and stropped his razor on it so rapidly that his hand went into a blur, while he stared at the customer.

The customer began to look alarmed.

"Watch," Patterson said to Juan. The blade went high and stopped for an instant, glinted, and flashed downward toward the customer's face. Patterson *slashed* the customer fourteen times on the face with blinding speed, while the customer's eyes darted all around, trying to track the blade. It was the world's fastest shave. If Patterson had miscalculated, the customer would have hosed the ceiling with spurting arterial blood. Patterson wiped his razor and snapped it shut. He said, "The face has fourteen parts. You can shave each part with one stroke."

The customer fearfully and slowly put his hand to his face, and his eyes went wide; his face was smooth and hairless.

As Juan progressed, he moved to the front of the room, where the lawyers of San Antonio sat in a row of chairs in front of big windows. The lawyers liked to gather in the front chairs, to talk with each other and to be seen in the windows. By the time Juan advanced to the lawyers, he had learned how to do flat-top crew cuts, known as flats. A flat required a heavy lubricant—butch wax. "One needed a good machine and a lot of grease to make the customer's hair stand up straight," he recalled. One also needed high artistry. The hair was greased with butch wax and pulled up vertically with a comb and buzzed back and forth with a pair of clippers until it was bristly, as smooth as a golf green, and stiff enough to rest a brick on. The wax was always jamming the clippers. Furthermore, if you lost control of the clippers, you could easily gouge a hole in the customer's flat, the way a golfer knocks a divot out of a driving green. Patterson would come over to see how Juan was doing. He would take Juan aside and say to him in a low voice, "These guys are lawyers. They want to look sharp, and so you have to get the blackheads out of their noses." After giving a lawyer a shampoo, a butch wax and a flat, a shave, and a mud pack, Juan would squeeze the lawyer's nose with both his thumbs to extract the blackheads, and then he'd proceed with the rest of the treatment: a jiggle with the electric machine to break up fat in the cheeks, a splash of rum aftershave. The lawyers said, "Thanks," and gave him thirty-five cents plus a tip. He graduated with a class-A barber's license in 1954.

In the Davis Mountains south of Pecos and Balmorhea, there is a cluster of white domes—the McDonald Observatory. A girl named Lily Dominguez worked there as a bookkeeper. Juan had known her since junior high school, and when he felt that he had a prospect of a career before him, he proposed marriage to her. She accepted, and Father Girán married them, but Apolonio did not fire his revolver at the wedding because the practice of *vivas* had run out of style with the modern times. Lily quit her work at the observatory in order to marry Juan. They moved to the city of Pecos, where Juan got a job in Angel's Barber Shop, on the edge of the barrio. In the barrio they called Angel "El Maestro." He came from Mexico, where the barbers have extremely refined manners. Angel could do anything—flats, ducktails, he could wax a mustache into points, he could make all the shapes and styles of sideburns, from Rudolph Valentino to Prince Albert, and he could shave heads. He had a row of showers in the back of his shop, for the use of which he charged twenty-five cents, and he bought a new car every year.

In the end, Pecos did not suit either Lily or Juan. Juan was not buying a new car every year. "I was very ambitious," Juan recalled. "I said to myself, 'How can a man get rich cutting hair?'" Meanwhile the observatory missed Lily. They urged her to return to the mountain and to bring her husband. They offered to train him as a night assistant. "I was pretty scared," Juan said. "What did I know about telescopes?" He and Lily moved to the McDonald Observatory, where the astronomers appointed him the night assistant on the eighty-two-inch McDonald reflector, then the largest telescope on the mountain, while Lily worked as the observatory's cook, housekeeper, and bookkeeper. The large McDonald reflector was a tricky telescope, and to operate it one needed the touch of a maestro. That was not all. The astronomers wore electrically heated flight suits (similar to the ones on Palomar) that were supplied with electricity by a wire that ran from a wall socket into the suit's rump. When the astronomers became overexcited, they would run around the dome shouting, unplugging themselves, dragging their wires all over the place. Juan learned that a part of a night assistant's job is to prevent astronomers from electrocuting themselves.

The same year he married Juan and Lily, Father Girán went to live in a retirement home for priests in New Mexico. Juan and Lily went up there to visit him. "He was very old then," Juan recalled, "but still very, very strong."

He was also delighted to hear that Juan had gone into astronomy. "My mind," Father Girán said, "doesn't remember things anymore. I would like to discuss astronomy with you, Juanito, but I find I have forgotten it. Yet I remember those nights . . . those nights when I told you about the stars. All those nights . . . You sit there with those astronomers. You listen to what they say. Now, you know, astronomers never get rich. But if you stay with them, you will learn, Juanito. Because astronomers are the chosen. They are the chosen." This was Father Girán's last conversation with Juan; he died shortly afterward.

After working at the McDonald Observatory for eight years, Juan and Lily turned their thoughts to southern California. In 1964, they moved to San Diego, where Juan found work in a physics laboratory, measuring the tracks of subatomic particles in a bubble chamber. By this time he had finished a correspondence course and had received a high-school diploma, and he had also begun to study computers. ("I am not totally ignorant of 'and/or' gates.") After several years at the physics laboratory he realized that he was about to be replaced by a computer, and so he decided to look for a job that would never be taken away by a machine. Night assistantship came appropriately to mind. One day he drove up Palomar Mountain and asked the mountain superintendent for a job. He reported to work on September 9, 1969, cutting underbrush and dusting the Hale Telescope with a mop. That was when he discovered his respect for heights, because he had to walk out onto the I-beams of the main tube, several stories above the mirror, pushing a dust mop in front of him. The mirror was covered, but if he had fallen on it, he still might have broken it, and he did not want to break that mirror, because while man is dispensable, telescopes are not.

By Christmas he had been promoted to relief night assistant. He learned how to slew the Hale Telescope—move it fast across the sky. Gary Tuton, who was then the senior night assistant, trained Juan. Juan was terrified that he might crash the Hale during a high-speed slew. What would it be like, he asked himself, to go down in history as the guy who had wrecked the largest telescope on earth? He reminded himself that he had never cut a customer.

The Hale had tendencies. It tended to pop its clutches. During heavy slews, one could smell burning rubber. The dome tended to lose track of where the telescope was pointing. Juan's notebooks tended to multiply. During cold weather, the film of oil under the horseshoe bearing might thicken, and the Hale would stop tracking the stars. That would trip an alarm buzzer, and the astronomer would open his mouth and prepare to scream. Juan would dash to a ladder resting against the telescope's north pier. He would climb three stories up the ladder until he came to a set of screws. He would take a little screwdriver out of his pocket and give one particular screw a quarter of a turn. One-point-one million pounds of telescope would resume tracking the stars, while the buzzer would shut off and the astronomer would return to normal.

Occasionally the telescope refused to slew. Juan would dive into the elevator and descend one level, then hurry along a walkway until he came to a bank of circuit breakers. Using both hands, he would hit every breaker switch as fast as he could, working his way down a wall of switches. If the Big Eye still would not move, he wrote down at least a dozen other things to remember. These included:

- Think of the oil pads.
- Check the clutch lights.
- Check the central hoist—if it is slightly raised, telescope will not slew.
- If the dome starts going back and forth, or "hunting," then the brake is too loose. A touchy little beast. Try just tapping the brake with a screwdriver first.
- If the mirror cover will not open and telescope will not slew, check the small DC motor next to the huge MG SET. Tap it if it won't start.

Sometimes the dome would get confused and start turning around and around, and the whole building would shake, and nothing could stop it. "Juan! This is crazy!" the astronomer would say.

"It's the phantom," Juan would say. Juan would dash into a room at the south end of the telescope. The phantom was in there. The phantom was a mechanical computer built by Bruce Rule. It caused the dome to follow—to phantom—the movements of the telescope. Juan would wiggle and snap moving parts in the computer. He would inspect and remove two vacuum tubes and wipe the tubes with a rag. Usually the dome would stop spinning.

The engineers seemed to be superstitious about oils. Cans and bottles had piled up on shelves next to Juan's locker. Some of the oils Juan learned the use of, and some he did not. There was Lubriguard Anti-Seize, Mobil Extreme Pressure oil, Graham transmission oil, Way Lube chain oil, Marvel Mystery oil, Gargoyle Grease. Bob Thicksten was trying to switch to a few simple oils, but Juan wondered if things like Gargoyle Grease and Marvel Mystery oil were not still ending up inside the Hale Telescope, either because those oils worked or because the engineers had bought them on special somewhere. Mobil had recently quit making Flying Horse telescope oil, which had caused observatories all over the United States to go into a panic. Thicksten had snapped up ten drums of Flying Horse, which he stored as if they were cases of '59 Margaux, in dark recesses of the dome. Thicksten believed that the use of Flying Horse oil was another of those black arts: perhaps unnecessary but foolproof.

After a nine-year apprenticeship Juan became the senior night assistant. Soon after his promotion the observatory installed computers to monitor and control some of the motions of the Hale Telescope. These digital systems had partly replaced Bruce Rule's clockwork computers. Juan now had a computer terminal at his control panel, through which he orchestrated many of the motions of the Hale. But the observatory had decided that no computer could ever be allowed to slew the Big Eye, because a computer would sooner or later crash the telescope during a high-speed slew. Man was not entirely dispensable. Especially this man, who could shave a wino with a straight razor and do a perfect flat. He had spent more time looking at the Hale's video screens than anybody, including Maarten Schmidt, which had forever changed his feelings about the earth's place in the creation. "Now when I look into the blue sky," he said, "I wonder where it all ends."

The Trojan planets were slow-moving, predictable bodies. Once you had sighted one, you could usually find it later. Carolyn Shoemaker felt no compulsion to quickly search the films that resulted from her and Gene's Trojan runs on Palomar Mountain. She did go over the films lightly, looking for fast-moving objects, and she found an asteroid called 1985 WA. 1985 WA could hit Jupiter, which was a pleasant surprise—it was a Jupiter-crosser. In any case, after New Year's of 1986, she and Gene returned to Palomar Mountain for a week, to survey the sky for earth-approaching asteroids, while the Trojan films remained in their glassine envelopes in Flagstaff, Arizona, unsearched for Trojan planets.

On Palomar Mountain, Carolyn found the first new comet of 1986. It turned out to be a regular visitor to the inner solar system—Periodic Comet Shoemaker 3. Then she found the second comet of 1986. This one, designated 1986b, was a long-period comet. It took a hairpin turn around the sun and left the inner solar system; it will be back in the spring of A.D. 2509. "I am now tied with Caroline Herschel," Carolyn said. "Of course, I am going to beat her." But the Trojan films remained in their envelopes, unsearched for Trojan planets. "As usual," she said, "I am sinking under a pile of films."

■ ■ ■

During March of that year, Periodic Comet Halley made an appearance in the sky. Maarten Schmidt's quasar team happened to be scanning on Palomar Mountain at the same time, and they were anxious to see Halley's comet, since none of them had seen it before. Maarten Schmidt brought his binoculars to the mountain. At four-thirty one morning, he and Gunn climbed to the catwalk. A cloudless wind came out of the west, hissing around the dome and snapping the tops of the cedars, warning of a storm coming from the Pacific Ocean.

Maarten held up his binoculars. "This is marvelous," he said.

"I had expected worse," said Jim.

"The comet looks fairly bright," Maarten said.

"This is quite spectacular."

Halley's comet had a white nucleus. It looked like a fuzzy star. The tail was cloudy and faint. The comet resembled a dust ball, a piece of celestial trash.

Maarten said, "Gad, these binoculars are *terrible*. I paid only fifty dollars for them. My apologies, James." He handed them to Gunn.

Gunn plugged the terrible binoculars into his eyes and turned the focus knob. His hair strayed in various directions.

In a philosophical sort of way, Maarten continued. "I think we are overdue, James, for a really bright comet."

"I absolutely agree."

"Like the great comet of 1843," Maarten said. "With a tail that goes halfway across the sky. It would be interesting to see what effect that would have on the popular mind, as it were. Whether we would hear talk of the end of the world and so forth." (Schmidt had studied comets with Jan Oort at the University of Leiden, where he had become interested in the effect of comets on the mind. He once remarked to me—we were talking about the Shoemakers and their comets—"I believe it was a comet in the seventeenth century that touched off a riot of shoemakers in Germany. There must have been more shoemakers then. And I bet you they were well organized.")

"The tail is developing nicely," Jim said. Then the binoculars strayed away from Halley's comet. He pivoted slowly on his feet, moving from point to point along the Milky Way. "The Trifid nebula is up," he said. "And there's M22—that's a fantastic globular cluster. And there's the Omega nebula." He handed the binoculars to me. He said, "You can see the Omega. It's right above the Lagoon nebula. What you can see with a pair of binoculars is actually quite mind-boggling."

The central bulge of the Milky Way fattened toward San Diego. The galactic core was rising over San Diego. Seen through a pair of binoculars, the central region of the galaxy is speckled with dots of glowing gas and cut with black streamers of entrained dust. Halley's comet hung immediately below the Milky Way—obviously near the earth. Halley's comet is a lump of black, dusty ice the size of Manhattan, shaped like a potato—and it travels on a chaotic earth-crossing orbit. It used to live out in the Oort cloud, until it felt the gravity of a passing star and fell toward the sun. Now it steams every time it approaches the sun. Someday it may break up and dissolve into dust, or it may become an extinct comet nucleus. If some kind of dark nodule remains, then one day Halley's comet may hit the earth as an asteroid, causing mass extinctions, although that is not likely. More likely, Halley's comet will have a close encounter with Jupiter during the next million years, and Jupiter will fling it off into the stars, to travel forever through the Milky Way.

Just three weeks before that, Carolyn Shoemaker had found an asteroid that she and Gene named Amun. Amun is the kind of earth-crosser that creeps up alongside the earth. Amun is a mile and a half across and is made of metal. "When Amun hits," Gene said, with a good deal of satisfaction, "it will make a *real* crater." Carolyn also spotted an object slanting through the plane of the solar system, moving fast—an object now designated 1986 EC. She measured its motion in two films taken as the moon was beginning to wax, washing out the night sky. Then she lost it. 1986 EC drowned in moonlight and has not been seen since.

"I hate to lose one of those things," Gene said.

Carolyn was chagrined, if not embarrassed. "I just hope it will be picked up someday," she said. Meanwhile the Trojan films remained untouched. "I'm not worried about those Trojans," she said. "Those planets are not going anywhere. They can wait."

On May 5, 1986, on Palomar Mountain, Carolyn found a strange minor planet in a pair of films that she and Gene had taken a few days before—a slow-moving asteroid, and going the wrong way,

against the flow of the Main Belt. Carolyn telephoned Brian Marsden, the director of the Minor Planet Center, and gave him a set
of rough coordinates. He gave the asteroid a temporary label: 1986
JK. The Shoemakers took more photographs of 1986 JK on following nights, while continuing to report the object's changing positions to the Minor Planet Center. Brian Marsden shortly informed
the Shoemakers and the rest of the world (via international astronomical telegram): "1986 JK appears to be an Apollo object
approaching the earth." Astronomers rushed for their telescopes.
The asteroid's slow motion was an illusion: the Shoemakers were
watching it head straight at the earth. JK reversed its apparent
motion, accelerated wildly, and boomed past the earth on June 1,
1986, at a distance of 2.6 million miles, one of the closest asteroidal
misses on record. By the standards of normal planetary motions
in the solar system, the apparition of 1986 JK was somewhat akin
to having the hair on one's head parted by a bullet. JK was traveling
on a long ellipse, like the orbit of a comet. Radio astronomers
using an antenna at the Goldstone station of NASA's Deep-Space
Network bounced a radar signal off the asteroid as it went by and
received a clear echo, which suggested that JK is a big object. Its
orbit takes it out as far as Jupiter. It is an earth-crossing Apollo
object, and a Mars-crosser, and a Jupiter-crosser. 1986 JK could
hit the earth, Mars, or Jupiter. At intervals of fourteen years, marking time like a metronome, it whips past the earth. Around Independence Day in the year 2000, it will be back in our skies on a close
pass.

During the autumn of 1986, one year after the Shoemakers had
taken the Trojan photographs, Carolyn decided that she would try
to find some Trojan asteroids. She selected thirty-three stereo pairs
of films to search. The photographs looked through the plane of
the solar system, into the Greek cloud of asteroids that traveled
ahead of Jupiter. Accustomed to looking for fast-moving asteroids
near the earth, Carolyn had to train herself to see slow-moving
specks hanging in space out near Jupiter. She set up her microscope
in a windowless storage room at the back of the library of the
United States Geological Survey's headquarters in Flagstaff, where
she could look at the sky all day without interruption. She fastened
pairs of negatives into the microscope.

She looked through binocular eyepieces and saw a white sky, as it is called. The photographs were negatives, in which the sky appears white and the stars black. (The human eye can more easily discriminate a black dot against a white background than the other way around.) She moved the films back and forth while stars passed through her field of view; this felt to her like flying through space. Each stereo pair of films contained about ten thousand dots— stars, galaxies, quasars, and asteroids. The stereo pairs had been photographed at time intervals of forty minutes. During forty minutes a Trojan planet would move against the background of fixed stars.

The stereo microscope was a clever tool for finding objects in motion—it made them pop into stereo. Trojan planets hung in the distance, barely in front of the stars. She could see the depth of the Trojan cloud; the Trojans were distant objects, well beyond the Main Belt asteroids that sprinkled the foreground of her films. An asteroid approaching the earth, on the other hand, could seem to hang in front of Main Belt asteroids. These earth-crossers moved aggressively, at odd angles. Traveling at perhaps thirty-four thousand miles per hour, an earth-crosser could go a long distance in forty minutes. "The asteroid is moving so fast," she said, "that you see a displacement in the image. You can't fuse it with your eyes— it seems to blur out. It hops."

She saw moving objects dotted all over her films; there is an almost unbelievable quantity of natural material floating around the solar system. Most of what she saw were Main Belt asteroids, many of them, no doubt, unknown objects. Comets were fuzz balls—a fog of gases flowed off them. Comets gave her a thrill. There was something about that misty haze that made her heart leap in her throat whenever she thought she might be looking at an undiscovered comet.

She kept her eyes moving, eight hours a day. "I try to move the film steadily across my field of view, without stopping," she said. "If you start paying too much attention to things, you start questioning everything: Is this an asteroid? Finding asteroids gets to be instinctive." When she saw what looked like a Trojan planet—a dot floating just in front of the stars—she marked it with a red pen. "These Trojans are fainties, kind of crummy-looking," she said. They resembled dust particles stuck to the surface of the film. She got into a habit of running her finger lightly over anything that looked like a Trojan planet, and sometimes a "Trojan planet" went away. She did not want to report the changing locations of a piece of dust to the Minor Planet Center. The stars formed trapezoids, rings, letters—her private alphabets. She would become intrigued with a galaxy. She would imagine herself as a traveler in space, approaching that galaxy. She explored globular star clusters. A globular cluster is a spherical cloud of several hundred thousand stars that orbits independently through the Milky Way. The sight of a globular cluster could cause a sharp detour through the Trojan cloud. At times she would forget what she was doing. She would come to her senses, realizing that she had become lost in a film and had no idea where in the film she was traveling. At other times she would wake up suddenly at home in bed—because if she spent enough time with the microscope, she began to find asteroids in her sleep.

After she had scanned and marked a set of films for likely Trojan asteroids, she made rough measurements of their positions, and then looked them up in a book that astronomers call the Russian Ephemeris. (The Russians keep track of asteroids for everyone else.) "I picked off most of the known Trojans that way," she said. She found plenty of Trojan planets, and they already had names.

The job consumed weeks of tedious effort. "I wouldn't work this hard if it wasn't so much fun," she said. In late November 1986, she found a speck that was not listed in any catalog, and it did not go away when she ran her finger over it. That same day she found another speck. She measured their coordinates using a machine called a measuring engine, and typed her measurements into a computer linked to the Minor Planet Center.

The next morning she received on her computer a message from Brian Marsden: CAROLYN: EXCELLENT! He informed her that she had sighted two unknown Trojan planets. He gave them the temporary designations 1985 TE3 and 1985 TF3. These planets had never been seen before. They were huge minor planets, each about fifty miles across—"elephants," as Gene called them. They were bigger than many of the moons of Jupiter, and, as Carolyn put it, "They are ours." Like all Trojan planets, they were at least as black as charcoal and probably blacker.

A week later she found another unknown Trojan, 1985 TG3. Then another, 1985 TL3. She also picked up a Trojan that a team of astronomers in Denmark had recently sighted, and she helped the Danish astronomers get a better fix on its orbit. She also recovered a Trojan—a Trojan that had been sighted by the Dutch astronomer E. J. van Houten in his original survey of the Trojan clouds but had been lost. In the end, she found four new Trojan planets and helped to secure identifications of two others. She said, "It's been an education for the eyeball."

Gene was pleased. He said, "The implication is that there is just a huge number of Trojans out there. We are only finding the elephants."

In order to give these planets names from *The Iliad*, the Shoemakers would probably have to wait for a year or longer, because an asteroid must be seen on three orbits around the sun before it becomes eligible for naming. Gene hoped that they would not have to scrape the barrel of *The Iliad*, but he was afraid that by now there might not be any heroes left.

The solar system remained a mysterious object to Gene Shoemaker. He had been wondering for a long time where these Trojan planets had come from. He had been wondering how they fit into the origin of the bunch of dust in which we lived. The common wisdom, he said, held that the Trojan planets were leftover pieces of the same material that had clumped together to form Jupiter, during the accretion of the planets. He had not often made a habit of buying common wisdom. "Lately," he said offhandedly, "I've been working to cobble together a theory. It would account for the formation of the planets Uranus and Neptune, the creation of the Oort comet cloud, the existence of these funny black objects near Jupiter, the late heavy bombardment of the moon, and the formation of the earth's oceans. If I ever get a week free, I'll write that one up."

He thought that he could explain the existence of water on earth and the existence of the Trojan planets all in one story. It was a hypothesis, a hunch about what had happened back then, when the solar system was coming together. He thought that the formation of the earth's seas might have been a part of an accretion process. Uranus and Neptune were large, icy planets, made of methane and water, with a rocky core. They had formed near the outer edge of the accretion disk that had become the solar system. During the era of fast planetary growth, the planetesimals—balls of silica, iron, tar, and ices orbiting the sun—had pounded together and welded to form planets. The accretion of the outer planets had not been a gentle process. As Uranus and Neptune grew, they experienced near misses with their own planetesimals. They played crack-the-whip with pieces of themselves; they whiplashed their planetesimals all over the place. Most of this debris was ejected into orbits beyond Pluto, where it is now called the Oort comet cloud. Gene felt that the Oort cloud was probably a haze of ice ejected from the solar system during the creation of the outer planets. Some of these wild pieces of Uranus and Neptune, rather than looping out to the Oort cloud, fell inward, past Jupiter. As they neared the sun (which had caught fire by then), they became dramatic comets.

Gene suspected that Jupiter had acted as a flytrap, perturbing these comets as they floated by, trapping them in the Trojan areas. Some force had to have decelerated these comets in order to trap them near Jupiter. Orbital specialists had not been able to come up with a mechanism that would slow down a large comet, but Gene had an idea: perhaps these comets had collided with small pieces of debris sitting in the Trojan areas. The comets had bumped into bits of rubble floating in the Trojan areas, stalled there, eventually burned out, and lost their tails.

He said, "My hunch—and this is Shoemaker's private view of the world—is that the Trojan planets are all extinct, captured comets." There might be nearly a quarter of a million black planets out there in the Trojan clouds, but according to Gene they represented "just a sniff of the total mass that was flowing through that region during the formation of the solar system." The planet Jupiter had thrown some of that mass into earth-crossing orbits.

With the naked eye one can see that something violent had once happened to the moon—the creation of the dark plains known as the lunar seas. To the naked eye they look like bruises. Galileo had thought they were oceans, but they are scars left over from what Gene calls the late heavy bombardment of the moon—immense impact structures bearing names such as the Ocean of Storms, the Lake of Death, the Lake of Dreams, Tranquillity Sea, and the Sea of Serenity, where Gene might have walked if he had only been eligible to be an astronaut. He believed that the lunar plains might have been made by stray chunks of Uranus and Neptune hitting the moon. With all that material floating around, there would have been a late heavy bombardment of the earth too.

"How do you go about fund-raising for the earth's oceans?" he had wondered. The standard theory for the origin of water on earth said that the water had come from volcanoes that had plumed water vapor into the atmosphere. "Essentially the standard theory said that water sweated out of the earth," he said. "We call that juvenile water." He thought that the best place to look for juvenile water was the sky. "You can envision one of these big comets smacking into the earth at twenty kilometers per second," he said. "The projectile would vaporize as it hit a rocky surface. A lot of the contents of the comet might rain out—you might get rains of water, carbon dioxide, and ammonia. The water would collect in the lowest basins. It would take about two hundred of the largest Trojan planets to make the ocean—biggies about two hundred kilometers across, Trojans the size of Hektor or Agamemnon. But there are bigger things than Hektor out there. Pluto, for example. Sure! The planet Pluto might well be a huge, extinct comet! If you could shake Pluto loose out of its orbit and bring it nearer the sun, you would have a *comet*, because the ice on it would start to steam away, forming a tremendous tail. One or two real big comets the size of Pluto, and maybe fifty Trojan-sized comets, and a bunch of itty-bitty stuff—that would deliver you the ocean. There's a problem with this theory. When I calculate the expected cratering rate of the earth during the late heavy bombardment, I get too damned much water!"

Comets probably contain hydrocarbons, and they may also contain traces of amino acids, the building blocks of protein. Gene did not think that amino acids could survive the heat of a large impact. But a small impact was another matter. He said, "Small cometary debris—there was plenty of it around during the late heavy bombardment—could be decelerated in the upper atmosphere and reach the surface of the earth intact." The earth's primeval seas might have contained a broth of water-soluble organic compounds, derived from steaming hunks of comets. The human body is 70 percent water, and much of the rest of it consists of organic, carbon-based molecules. To Gene Shoemaker it seemed not impossible that the human body might be largely former comet.

The Trojan planets, in the Shoemaker view of the universe, were extinct cometoids covered with a goo of carbonaceous dust or tar. What you might have out there, Gene thought, was an asteroid belt made up of nearly a quarter of a million ancient comets. "The existence of large Trojan clouds," he said, "would amount to circumstantial evidence for a huge flux of comets early in the history of the solar system. I think that these Trojan planets are preserved samples of the same guys that were delivering the oceans to the earth. The point of this is that there's a whole other asteroid belt out there near Jupiter still to be explored."

D on Schneider's office at the Institute for Advanced Study, in Princeton, New Jersey, looked over meadows to deep woods, which were beginning to incandesce from autumn frost. Don was sitting at a table that held two computer screens, two keyboards, and a video monitor. A photograph of the Heartwell farm hung on the wall. The time had come for an attempt to mine the sky for quasars. His lifework, or so he claimed, consisted of about two hundred reels of computer tape and a program named Cassandra. The tapes contained electronically recorded images of some of the things out there—of cataclysmic stars, of quasars, of packs of feeding cannibal galaxies, of gravitational lenses, and of the tapestries of sky generated by Maarten Schmidt's search for the redshift cutoff. Cassandra was an image-processing computer program. She could recognize patterns.

"Cassandra," he said, "has been searching a crowd of faces. She has been looking for interesting faces." He hit a key, and an image of sky came up on the screen, showing galaxies and stars with candle flames emerging from them—smears of spectral light. The screen displayed a moment from a night's transit—a piece of sky that happened to be located within the Trojan cloud of planets. The screen was painted with many spectra—the broken light of about a hundred stars and galaxies, and possibly a Trojan asteroid or two. (Cassandra had no way of recognizing Trojan planets.) Don turned to a pile of papers and studied a jagged line plotted on paper—a line describing the peaks and valleys of color in one particular object on the screen. He studied the paper. He peered at a candle flame on the screen. "Ooo," he said, "that looks promising."

Cassandra had identified this thing as a possible quasar.

He put his face closer to the screen and studied the object. "Nope, that's an M star," he said. "A red star. Cassandra thinks its a quasar." He marked an X on his paper. He hit a key, and a new image of sky came up.

Out of a tapestry of sky containing about 120,000 stars and galaxies, Cassandra had picked out a list of about 2,000 quasar candidates—objects showing bands of bright color resembling the light of quasars. Cassandra tended to discover things that were not quasars: stars saturated with metals, galaxies with glowing cores, and glitches in the data. Don had to check Cassandra's discoveries by eye, to weed out the false alarms. Then he and Schmidt and Gunn planned to return to the Hale Telescope to take detailed spectra of the remaining candidates. They hoped that a handful of these candidates would turn out to be quasars. Given luck, one or two of those quasars might turn out to be extremely distant, deeply redshifted monsters.

He had been keeping two VAX computers running simultaneously around the clock for weeks, eating galaxies by the megabyte. He said, "I'm just a bit player here. No pun intended." He did not want a single quasar to slip through his nets. Maarten Schmidt wanted nothing less than a perfect quasar filter. "Maarten Schmidt casts a long, thin shadow," he said. If Cassandra missed any quasars, then the search for the edge of the universe would end in New Jersey—"And I'll be heading south of the border," he remarked.

He offered to introduce me to Cassandra. She could talk a little. He sat down at another of his computer terminals and hit a few keys.

Cassandra said, on the screen, MAY I HAVE YOUR LAST NAME, SIR? "Schneider."

HELLO, MASTER!! I HOPE I PERFORM SATISFACTORILY.

Don said, "She calls me Master. Other people she calls Junior. Give her your name."

I typed, PRESTON.

GREETINGS! I'VE BEEN WARNED ABOUT THE PRESTONS.

He said that Cassandra had previously been notified of my arrival. "But if the program doesn't know your name, it dies. It says, 'That is no way to address a lady.'"

Cassandra could do all kinds of things. She could do an "auto object grunge," in which she measured the location of every star within a crowd of stars. She could transform herself into a hunter / seeker, looking for the exact center of a galaxy. She could plot the colors of light from a quasar into a jagged line, like a stock-market chart. She could also construct artificial stars and galaxies, for testing purposes.

Don said, "Let's play God," and typed a few commands to Cassandra.

She said, FORMING SKY.

A night sky appeared on a nearby screen, speckled with stars. An imaginary sky created by the computer.

She said, I AM BUILDING A STAR.

A bright star appeared on the screen.

I AM BUILDING A GALAXY.

Nothing happened.

"What's going on?" Don muttered.

A galaxy appeared.

"Ah! There we are," he said. Typing fast, he told Cassandra to make a globular cluster of stars.

FORMING SKY.

Nothing happened.

We waited.

Nothing continued to happen.

"Uh-oh," he said. "It's making a hundred thousand stars. That will take days." He told her to try something less ambitious.

After a while a shotgun blast appeared on the screen—a globular cluster—and she said, I HAVE CREATED 200 OBJECTS.

Cassandra contained something like fifty thousand lines of code. "I don't know the exact number," he said, "because it's always changing." He had chosen the name Cassandra for a reason. Cassandra, he explained, had been the daughter of the king of Troy. The god Apollo had fallen in love with her and had given her the gift of prophecy. But when Cassandra had refused Apollo's advances, he had cursed her and told her that her prophecies would come true but that nobody would believe them. During the Trojan War, when the Greeks had left a wooden horse by the gates of Troy, Cassandra had warned her fellow Trojans of the danger. They had ignored Cassandra. "My program works," he said, "but nobody believes it."

A large program such as Cassandra can often be passed around during its development. The program had been started by one Robert Deverill, who had given it to Don Schneider and to a Caltech classmate of Don's named Peter J. Young. P. J., they called him. Don and P. J. had both entered Caltech in 1976 as graduate students in astronomy. They were the astronomy class that year— a class of two. P. J. Young was a thin Englishman with a rapid way of talking and a quick sense of humor. He was considered to be the most brilliant astronomer that Caltech had produced in many years. Don and P. J. collaborated in developing the Cassandra program. But Peter J. Young was a victim of deep depressions, which he covered up—no one knew about them. One day not long after he graduated, P. J. Young shut the door of his office at Caltech and killed himself. He had been one of Don's best friends. The loss left Don with an orphan, the computer program, whom Don had continued to raise and educate. Now he was the only person on earth who understood Cassandra.

Cassandra would soon be hunting / seeking through pictures taken by the Hubble Space Telescope. She had come a long way, but P. J.'s death had left Cassandra with a dilacerated heart. The pain would never go away, but Cassandra hid it fairly well now, although every time her master closed the program, she asked him, DO YOU REALLY WISH TO LEAVE ME? and if he answered yes, then she said, LIVE LONG AND PROSPER, MASTER.

He did not think, however, that computers would ever have a self-awareness akin to human consciousness. He said, "I think that the mind's capacity to be aware of itself is equivalent to what you would call the soul. If we ever build a machine that is self-aware, then I will be worried. And yet we are so tiny. Sometimes it just amazes me that we can understand the structure of the universe at all."

He said that he tried to avoid discussions with people who believed that the universe had been created six thousand years ago, and who argued that God had made the universe *appear* as though it were billions of years old. He said, "We all have to deal with God in one way or another, even by saying that God does not exist.

I am nobody to say who God is, but I feel that God is not dishonest. It is possible that we were created five minutes ago, along with our memories. It is also possible that the universe was created six thousand years ago, along with an apparent history. There is no way you can refute that. But it requires a dishonest God. I would not want a God who would make us in His own image to be dishonest." All the sky lay before astronomers, an open book displaying pages going back into history. As one looked into the age of quasars, one could read the beginning of the story. One could see a chronicle of immense time inscribed in the text of light. Perhaps God had made the sky as an illusion, like a projection in a movie theater, but Don preferred to believe in a God who had made a four-dimensional sky that displayed time as well as space, and who had saturated the sky with forces that over a hundred million centuries had given rise to the sun and the earth and to men and women with eyes to look back and discover eternity.

▪ ▪ ▪

"In the early days of a galaxy," as Jim Gunn expressed it, "there must be copious gas floating around." The universe, during the first billion years of its childhood, was crowded with matter. Clouds of hydrogen filled space, mixed with early generations of stars, gathering into galaxies. "This gas," Gunn said, "has a tendency to cool. As it cools, it sinks into the center of the galaxy. There is no way the gas can get out of the galaxy once it cools off. The gas has to form a big condensed object in the center of the galaxy. The gas is moving. It has some kind of random motion. When it collapses, it goes *zoop*, down into a rotating disk. The disk tries to make a star, of course. But it fails, because it has too much mass. So it's gonna make a black hole. That is its only fate."

Pierre Simon, the Marquis de Laplace, originally proposed the existence of such an object, in 1796. Laplace imagined that if an object were heavy enough, its gravitational field would curl around the object and would prevent the escape of any light from its surface. The object would drown in darkness. He called it a *corps obscur*, a lightless body. Modern astronomers call it a black hole.

The engine of a quasar appears to be a runaway accretion disk swirling around a black hole. That engine may be no larger than our solar system, which is a microscopic point in comparison to the size of a galaxy. Yet a garden-variety quasar burns one hundred times brighter than a galaxy. A garden-variety quasar emits the light of one trillion suns. There are quasars superbly brighter than that. An ultraluminous quasar in the constellation Cepheus shines sixty thousand times brighter than a galaxy. This quasar is close to the beginning of lookback time, and yet it can be photographed with a ten-inch amateur telescope, because it shines with the light of one quadrillion suns. A number like one quadrillion is actually inconceivable, but I shall try to explain it this way: one quadrillion grains of sand would fill a line of dump trucks five miles long.

There is reason to believe that a quasar is an extremely massive object. Albert Einstein's simple expression, $E = mc^2$, states that energy and mass are interchangeable; that energy can become mass and mass can become energy. A nuclear bomb is a primitive mass-converter. When a bomb called the Fat Man went off over Nagasaki, it transmuted an amount of plutonium having the weight of two peanuts into kinetic energy and light. An object giving off the light of a trillion suns must be somehow connected to a tremendous mass, and it must be converting much mass into energy.

There is another reason to believe that a quasar is a supermassive object, something much heavier than a star. Photons of light striking matter exert a faint pressure—the pressure of light. A person standing in direct sunlight receives a pressure from sunlight equal to about three tenths of a milligram, or the weight of an ant's thorax. The explosive pressure of light flowing out of a quasar would be enough to rip any star apart. A quasar must be bound together under enormous gravity. Otherwise it would puff up and rupture with out-streaming light. Astronomers have calculated that the flow of light from a quasar must be counterbalanced by the gravity of at least one hundred million suns, or else the quasar would balloon into a cloud of gas and disappear. A quasar is a detonating photon bomb that refuses to blast itself into oblivion.

If a mass equal to one hundred million suns were concentrated into an area the size of a solar system, that much mass would puncture a hole in spacetime. It would create a black hole and fall down the hole. No light can escape from a black hole. Any object that happens to fall into a black hole—a shoe, a star—is accelerated to the speed of light as it falls in. It disappears at the surface of the black hole. Between the fifth of February, 1963, (when Maarten Schmidt first perceived the immense energy in a quasar) and today, a general opinion among astronomers has emerged, that a quasar contains a black hole—not that any astronomer has ever seen a black hole, either inside a quasar or anywhere else, but a black hole seems to be the only object that can account for the light of a quasar.

At the center of a youthful galaxy, a cloud flattens into a wheel of gas, dust, and (perhaps) planetesimals—an accretion disk. At the hub of the disk, a giant protostar undergoes nuclear ignition. The protostar feeds on inwardly falling gas until it grows too heavy to support itself. The poles of the star suddenly collapse, and the star turns into a spinning doughnut. The hole in the doughnut falls out of the universe—it implodes, collapses in space, collapses in time, acquires great entropy, and redshifts to infinity. It becomes a black pinpoint, a black hole. The spinning doughnut-star sheds its inner rim into its own black hole. The doughnut eats itself. The surrounding accretion disk—gas, dust, planetesimals, whatever— draws inward and swirls around the hole. The disk spins faster, heating up through friction. More and more gas arrives in the disk and tries to flow down the hole. This is a solar system that went down a rat hole in spacetime, sucking a cataract of matter after it.

The disk spins so rapidly that the material in the disk has trouble getting into the hole. The disk glows from turbulence and friction. Matter spills from the inner lip of the disk into the black hole, putting some torque on the hole. The hole spins up until it is rotating at a velocity known as the extreme Kerr solution— approaching the speed of light. The hole screws space and time around itself and tugs at the inner lip of the accretion disk. Magnetic fields whip through the disk. The disk fattens and begins to shine in unspeakable colors—the theorists toss around phrases such as "upscattered soft photons" and "synchrotron radiation"— but nobody really knows what the colors of a burning accretion disk might be. Gunn said, "The disk progresses slowly into the black hole, and friction inside the disk is what gives rise to the fireworks." A hydrogen bomb converts about 0.7 percent of its core mass into radiant energy. A burning accretion disk can liberate up to one third of its mass into out-streaming light as the remainder of the mass gargles down the drain.

A black hole is a very large amount of mass crammed into a very small place. If the earth, for example, were to be forced into gravitational collapse, it would make a black hole the size of a golf ball. If a black hole the size of a tomato were placed into a low orbit around the earth, its gravitational pull would drag oceanic tides over the continents. It would crack up the continental crust, triggering volcanic eruptions. The earth would go into a binary orbit with the tomato. The tomato would pull the earth into a teardrop. If the hole and the earth touched, the hole would go into orbit *inside* the earth. It would spiral toward the center of the earth, devouring matter. The earth would melt, vaporize, emit X rays, and churn down the black hole. After the tomato had eaten the earth, the tomato would be a little bit fatter, tending toward a Burpee's Big Boy. A hole that has swallowed a hundred million suns would fill the orbit of Mars. The brightest quasars may contain a hole that has eaten several billion suns, which would fill the orbit of Pluto.

Around this hole the accretion disk would be burning brilliantly, but the disk might extend outward a great distance, perhaps as far as a light-year, dimming off gradually. At its outer limits the accretion disk might merge imperceptibly into the disk of the galaxy itself and sparkle with newly formed stars, which had condensed out of the disk as it spiraled toward the hole. "If you immerse one of these monsters in gas, as at the center of a galaxy," Gunn said, "it is going to grow. As these things get bigger, they become brighter. And more voracious. There's a common misperception, though, that black holes have to eat everything in a galaxy. Most of the matter in a spiral galaxy can't get near the galaxy's central black hole, because it is rotating around the center of the galaxy—as we are. There are natural limits to the growth of a black hole. Pretty soon the black hole runs out of food. It starves. The rise of quasars is the growth and saturation of a monster. The monster's food supply gets cut off, and the decay of the quasars chronicles the monster's starvation." These starved objects would still sit in the centers of some galaxies. They are not feeding anymore, or they are feeding only sporadically. The core of the Milky Way may contain a modest black hole that has swallowed the mass of a million suns (not enough mass ever to have caused the Milky Way to go quasar).

Astronomers are sure about only one thing regarding quasars: they are sure that they do not understand quasars. Bohdan Paczýnski, an astrophysicist who, like many, has spent a portion of his career contemplating the enigma of quasars, had this to say: "Our understanding of these accretion disks is comparable to astronomers' understanding of stars before the discovery of nuclear fusion." Jim Gunn expressed it this way: "We don't actually know that quasars have anything to do with black holes at all." The only bona fide accretion disk that astronomers have ever seen through a telescope is the rings around Saturn.

But the theorists enjoy trying to imagine what kind of an object makes the colors of a quasar. If a giant accretion disk—the core of a bright quasar—was located in our sky at the distance of the nearby star Alpha Centauri, the hot, central part of the disk would appear to be about the size of a penny seen from one hundred yards away. The light coming out of that penny would be two hundred times brighter than the sun. If you went outdoors and tried to look at it, your hair and clothes would burst into flames and your skin would char. You would absorb a lethal dose of gamma rays and X rays. If the quasar was emitting plenty of microwaves, as some quasars do, you would get a dose of radar that might cause you to boil internally. The experience would be not unlike standing close to a nuclear fireball at the moment of ignition. If you could somehow look at a quasar from nearby, through a glass darkly, you might not see the accretion disk. The disk might be enveloped in a shining globe, filling a great portion of the sky—a corona of hydrogen gas, perhaps one light-year across. The pressure of light coming out of the accretion disk would support the corona in a delicate stasis against collapse, preventing the corona from shrinking into the hole at the center of the quasar. The corona might be threaded with tendrils of fast-moving hydrogen, emitting Lyman alpha light.

Space around the quasar would be packed with stars, because the center of a galaxy is a well of stars. The stars would be orbiting the quasar, feeling the gravity of the black hole. Stars would pass through the quasar's corona and through the accretion disk inside the corona. Most stars would come out intact, except for the red giant stars. If a red giant passed through the hot part of the accretion disk, the giant would get a haircut and come out as a white dwarf. Quasars have been known to let off great pulses of light—implying that a cloud of gas hit the accretion disk and shirred into it, which would cause the disk to spike hundreds of times in brightness, which might make the quasar's corona puff and horripilate. The brightest quasars—the ultraluminous quasars—may be eating a steady diet of gas, each year, equal to the mass of a hundred stars the size of the sun. If the earth was vaporized and injected into an ultraluminous quasar, its total mass would power the quasar for one second.

You might see an opposed pair of jets streaming out of the quasar in opposite directions—the fearful symmetry of the particle beams. These are fountains of electrons and other subatomic particles that are thought to squirt from the poles of some spinning black holes. They can mushroom into clouds of gas millions of light-years across, large enough to swallow Local Groups of galaxies. The jet that Maarten Schmidt saw drilling out of his first quasar—the quasar called 3C 273—is apparently a particle beam as long as three galaxies.

▪ ▪ ▪

The tapestries in Maarten Schmidt's house made Don Schneider feel strange. They had been woven by Maarten's wife, Corrie. She had titled one of them "3C 273." It was a swirling disk, five feet across, knotted with gobs of material, and at the center Corrie had mounted a glass photographic plate of the quasar and its jet, taken on the Hale Telescope. Don now sat in the Schmidts' living room, telling Maarten that he had dug seventy-three good quasar candidates out of a strip of sky. Maarten and Don discussed the possibilities. Maarten hoped that some of these candidates would turn out to be quasars near the edge of the known universe. "You can be cheated by statistics," Maarten allowed, "just the way people can win or lose at Las Vegas. I'm knocking on wood right now. I don't want to knock wood too hard for fear that the weather or the statistics will turn against us." Don and Corrie and Maarten roasted shish kebab in the backyard, watching the sun go down. The Principal Investigator was a little nervous because he had been waiting for a year and a half to see what kind of crop of quasars these strips of sky would yield.

The next afternoon, Maarten and Don drove in Maarten's brown Ford to the summit of Palomar Mountain. They found Jim Gunn in 4-shooter's garage, surrounded by pieces of 4-shooter. They bolted into the Hale Telescope a sensitive CCD instrument, built by J. Beverley Oke, known as the Double Spectrograph.

Jim left off tinkering with 4-shooter long enough to look at some sheets of paper that Don had brought with him and which contained jagged lines—crude, almost illegible spectra of the candidates. Jim weeded out a handful of unusual spectra. These objects, he thought, might be deeply redshifted quasars. Maarten and Don both claimed to be skeptical. At the close of twilight they fed the coordinates of the first candidate into the telescope.

On the television screens in the data room, something that looked like a star came up. They took a spectrum of it. It was a star. A dud.

They slewed the telescope to the next candidate. They saw an anonymous galaxy with a hot nucleus. They put the slit of the spectrograph across the nucleus of the galaxy and decomposed its light, to see what kind of a creature this was. They waited for the computer to digest the light, and then a jagged line appeared on the screen, a line describing the intensities of color in this galaxy. They studied the line, reading the text of the light: it was a Seyfert galaxy—a spiral galaxy with a miniquasar burning in its core. This miniquasar was irradiating unto death, for all anybody knows, a few million solar systems near the center of the galaxy. Nothing fancy.

At eight o'clock in the evening they put the slit of the spectrograph over the first of Jim's deep quasar suspects, a point of light resembling a star in the constellation Aquarius. They opened the camera and gathered light for fifteen minutes. A spectrum came up on the screen, displaying the object's broken light. They saw glowing hydrogen and carbon. The thing was obviously a quasar.

The quasar was a monster. It showed a Lyman alpha emission line. Ordinarily this line would be ultraviolet in color and therefore invisible, but here it had been redshifted down into green. This quasar was receding at close to the speed of light, carried along in the Hubble flow, the expansion of the universe. It was a deeply redshifted quasar.

The astronomers continued to break up the light of candidate objects. They found another Seyfert galaxy. They found an N galaxy—a galaxy with a blue, starlike center. They complained about the seeing—the air over Palomar was rippling tonight. At eleven o'clock they pointed the Hale Telescope at another of Jim's deep suspects, a starlike object in the constellation Cetus.

Jim studied Don's paper. He said, "I think the redshift on this thing will be three-point-eight."

They opened the shutter and collected the object's light for half an hour. They closed the shutter and called up the spectrum. A ragged, up-and-down line appeared on the screen. The line resembled the silhouette of a conifer forest.

Maarten Schmidt had never quite gotten over his surprise at what a change the computers had wrought. He could read the text of light on a screen. This quasar had a stunning Lyman alpha peak, a spike of color that ordinarily would be ultraviolet, but this one had been dragged down into the color yellow. Now *there* was a redshift. The peak was lacerated with fine absorption lines—razor cuts in the spectrum, betraying clouds of invisible hydrogen gas in front of the quasar, perhaps swirling around the quasar. He saw ionized silicon in the quasar. He saw oxygen. He saw nitrogen. He saw the same elements that human bodies and the earth are made of, except that he was looking into the early universe. The photons just gathered in the Hale mirror had been possibly older than the Milky Way.

Punching buttons on a calculator, Maarten estimated that this quasar had a redshift of 3.8, or 380 percent. Gunn had been right. Later the astronomers would give the quasar a name: PC 0131 + 0120. "I suppose we could have named it after one of Maarten's daughters," Don would remark, "but Maarten has discovered too many quasars, and he only has three daughters."

By the end of the next night the astronomers had found five more quasars with deep redshifts. "A couple of years ago," Don remarked, "I would have gone nuts finding just one of these things. Familiarity breeds contempt."

The following night Gunn rebuilt 4-shooter in the garage, working by flashlight. Juan Carrasco noticed a weather front coming in. Maarten spent much time on the catwalk, watching clouds gather. He saw lightning on the horizon, which bothered him. Lightning can contaminate sensors and make it difficult to read the colors of a quasar. As the weather deteriorated, they managed to get another remote quasar. A few minutes before dawn, they shot the last deep suspect, which turned out to be another quasar near the beginning of lookback time. The sun came up, and clouds enveloped the mountain. The team had now turned up nine high-redshift quasars, all of them at the limits of the optically known universe.

Gunn finished his work with 4-shooter and left for the East Coast. In the afternoon a thick fog covered the mountain. Maarten and Don kept a vigil in the dome, even though the weather was looking poorer all the time. Maarten hunched over a desk, elbows out, in a pool of lamplight. He was wearing his brilliant red shirt. When Bach came on the stereo, he whistled, but he did not whistle for Beethoven. He was sketching something on a piece of yellow graph paper.

Don stretched out in a chair and tried to read a newspaper. He kept dozing off. ("These older astronomers have a lot more stamina than I do," he said.) He went up on the catwalk to try to wake up, and he saw that the fog had become bursts of rain. When he returned to the data room, Maarten said, "Come over here, Don. I want to show you something. Look at this." Maarten placed the paper under the lamp, then said, "I have been fitting our earlier data into these quasars we have just found." He removed his glasses, took up a pencil, and squinted. "I can put in the last points," he said, and then he traced the shape of the redshift cutoff.

"Whoa," Don said.

Maarten Schmidt plotted the space density of the quasars over the lifetime of the universe. He started the line at Time Zero. At Time Zero—the creation—there had been no quasars. His pencil traveled horizontally for a while, indicating the passage of the dark time, when apparently no quasars had existed. For roughly one billion years the universe had remained relatively dark while it exfolded in secret. Then Maarten's pencil line moved upward. The earliest quasars burst across the face of the deep, here and there, each one spreading the luminosity of trillions of suns—gamma rays, X rays, optical light, heat, microwaves, radio—beacons of unearthly color. These were the primeval quasars: the elusive early population, a fossil bed that Maarten had wondered if he might find. Quasars had been scattered thinly at first. Then, during a brief interval of time, Maarten's pencil line went to the top of the graph—the quasar population exploded and ascended into brilliance. Somewhere around two billion years after the creation, the universe had dazzled with quasars. Maarten's line crested laterally and started downward in a glide. The quasars peaked out rather slowly. One billion years passed. Maarten's line dropped faster. The population of quasars collapsed. The line flattened; the quasar population thinned out. Five billion years passed, then ten billion years, and when Maarten's pencil touched Time Present, the quasars had ceased to exist.

Don watched the chronicle of first light, feeling the hair on his neck crawl. He had not expected Maarten to try this, even though somewhere in the back of his mind he had known all along that these nine quasars might imply the shape of the redshift cutoff. Something like the following thought crossed his mind: This is the first glimpse.

Maarten dropped the paper into his briefcase. He felt slightly embarrassed. He began to think that he might have gone too far in drawing the picture. The curve might be wrong. But he could not help showing his sketch to Don. "I was boasting to him," Maarten would later admit. "Having no high-redshift quasars didn't tell us much. Once you get some, it is really tempting to make assumptions."

They walked down to dinner at the Monastery. When they returned to the dome, Maarten suggested a game of pool. The pool room is a chamber on a lower floor. There, in the old days, on rainy nights, the nabobs among the Palomar astronomers had hunkered down and gambled for nickels. Maarten and Don shot cowboy pool, a game that Milton Humason brought to Palomar, and the

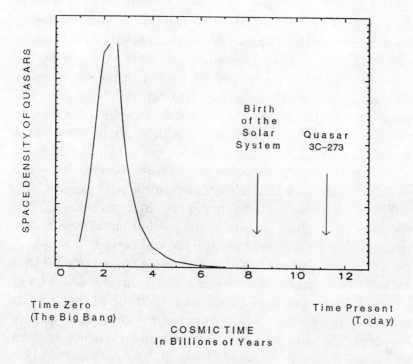

The rise and fall of the quasars, as seen by Maarten Schmidt. This curve shows the number of quasars born and burning over time, as the universe evolved. You can see that quasars started to burn when the universe was just a few hundred million years old. After two billion years the universe was rich with quasars, and then they quickly faded away. Now they've gone dead, as the graph shows. The quasar called 3C 273 is shown. It's actually quite close to us in time and space, and yet it seemed a very long way off when Maarten Schmidt first identified its distance, in 1963. (Graph courtesy of Maarten Schmidt, from an article by him in the *Journal of the Royal Astronomical Society of Canada,* 1993)

only pool game played there by the older astronomers. You rack and break only three balls, and the game switches to billiards at the end. Maarten told Don about his and Corrie's plans for travel. They did not talk about quasars. On Palomar Mountain it is considered bad form to talk science during a game of cowboy pool.

For his part, Don sensed—or believed—that the astronomical community had become concerned that Schmidt, Schneider, and Gunn were not finding any distant quasars. Now he could imagine people saying, "These guys finally got their act together." Don would later say, "Finding those quasars certainly made the years of frustration worth it. I shouldn't call it frustration. Nature is the way she is." He took hard, sloppy shots at the pool game.

"Our work is finally starting to pay off," Maarten believed. "We now have a simplistic picture of the redshift cutoff. We have a prospect that we can come up with the real curve in a couple of years. The assumptions in the little drawing that I made for Don are what we are going to investigate." Nine quasars added up to a hint. A larger sample of quasars might tell the whole story. "It will be of interest to see just how fast the turn-on of quasars happened," he said. "It seems that the turn-on was not as sharp as we had originally supposed. That will have to come out of a good, solid statistical sample." He suspected that a quasar might be the birth of the nucleus of a galaxy. If so, then the quasars would be beacons signaling some kind of birth-wave of galaxies. "Quasars may happen in many galaxies," he said, "but that we don't know. It depends on the duty cycle of a quasar—how long an individual quasar lasts. We don't know how long a quasar lasts."

His graph had showed the growth and retreat of a forest. A forest could last much longer than a tree. Trees grew and died, the forest waxed or waned. If the core of a galaxy were to go quasar, how long would the blow-off last? Would the galactic core go off like a flashbulb? Or would the quasar burn steadily for a billion years? "If the life span of an individual quasar is brief," Maarten went on, "then almost every galaxy could have its own quasar. If quasars last a long time, then not every galaxy can have its own quasar." There were other mysteries. Had the quasars been hidden behind walls of dust in the early universe? Did a quasar begin its existence shrouded in a dust cocoon? Did quasars hatch out of cocoons? Or had the early universe been free of dust, and had the quasars simply brightened rapidly? It seemed that the Hale Telescope had now been pushed as far as it would ever go, at least in sheer distance. As to whether the Hubble Space Telescope might detect any structures or objects beyond the apparent redshift cutoff, nobody could say.

Now that he could chart the redshift cutoff, he wondered if mapping the far side of an ocean was as important as discovering the ocean in the first place, as he had done two decades ago.

His feeling brought to mind the words of Robert Louis Stevenson: "It is better to travel hopefully than to arrive." When the game switched to billiards at the end, he saw that he was far ahead of Don Schneider. He tried to take some goofy shots to let Don catch up, but he accidentally won the game. Maarten was a reluctantly precise billiard player.

That characteristic fitted the Maarten Schmidt of Don Schneider's imagination—a European gentleman who had appeared in *Reader's Digest* at Don's grandmother's house when Don was eleven years old. When he considered the workings of time, it seemed as though he were living through the happy ending of a book—shooting cowboy pool with Maarten Schmidt, they having become the first two people on earth to glimpse the rise of the quasars. It seemed as believable as life itself.

For Maarten Schmidt, the best moment of the experiment had actually occurred long ago, when he had first suggested to Jim Gunn that they might throw a net for quasars by scanning a CCD camera across the universe. "That was probably my happiest moment in this work," he remembered. He had never imagined himself as what he called "technically developed," and yet he had taught a trick to James E. Gunn. That gave him a greater feeling of satisfaction than having seen the redshift cutoff.

The Flemish weavers put their stitches into the back side of the cloth. They could not see the true form of their work until they stopped weaving and walked around to the front of the tapestry. Don wondered what people of the future might say about the birth of the quasars. He said, "A hundred years from now, for all we know, somebody may say, 'These people had it all wrong.' Nature is always doing something under the rug where you can't see it.

Every generation thinks they have uncovered the last rug. They lift up the rug, and they find another rug underneath."

▪ ▪ ▪

Maarten Schmidt hungered for quasars, and the scanning went on. One night at the Big Eye he walked back and forth in the data room, humming to himself. He fiddled with the contrast on the video screen. "Isn't that a nice spiral," he said, touching a drifting galaxy. "Whack," he said, and gave the galaxy a twist with his finger.

"Oww!" said James E. Gunn, stirring in a chair.

"What's up, James?"

Gunn glanced at the clock. "It's four-thirty! Never mind me, Maarten. I can punch a return key with the best of them." Gunn stood up. He said to me, "You said you wanted to look at the mirror."

"What's going on?" wondered Don Schneider.

Pulling on his down jacket, Gunn said, "I thought I would take Richard Preston up to the top of the telescope and let him look down on the mirror."

Don smiled skeptically. "What are you trying to do, Gunn, get a spectrum of Richard's head?"

I followed Gunn inside the dome. He led the way onto the aluminum diving board, the lift that carries people to the top of the telescope. The lift wobbled under the weight of two people. He turned off his flashlight. He punched a button. There was a rumble, and the lift went up along the inside of the Hale dome, while the yoke and horseshoe bearing dropped away into darkness. The lift stopped. We were hovering at the top of the dome, beside the lip of the telescope, and the stars hung close overhead.

I fumbled with my notebook and pencils. Gunn reminded me not to drop any pencils when I looked down the tube: a pencil falling five stories through the telescope could damage the mirror.

I walked to the end of the lift, leaned over a railing, and looked into the barrel of the Hale. Then I realized why Gunn had warned me not to drop any pencils. A sheet of stars was hanging a few feet from my face, floating in the mouth of the Hale Telescope. It was an optical illusion. The stars seemed to have been thrown to the top of the telescope, where they were suspended in space.

I waved my hand through a veil of stars. The illusion was perfect. It seemed as though if one reached out a hand, one could catch a fistful of stars from the mouth of the Hale. The mirror, far below, seemed to be a flat black void.

"Nice, isn't it?" Gunn remarked. "The mirror is projecting an image of the sky into your eyes. Those are real images of stars, as far as your eyes are concerned. Your eyes are cameras. The eye is always a camera, of course, only now it's a camera on the Hale Telescope."

The Hale had magnified the stars and their colors, chryselephantine colors—white, blue-white, and pale gold. At that moment I felt an envy of all astronomers, as well as an envy of the builders of an instrument that could enclose and reimage the creation.

"Hah, what a fantastic night," Gunn said. "Arcturus is absolutely steady." He leaned back, put his elbows on the rail of the lift, and looked up at the sky. His glasses glinted faintly in the starlight. "Astronomy is not terribly important," he said. He fell silent for a moment, admiring Arcturus. Then he said, "Although it is one of the more important things we do as a species." He did not see any contradiction there. He said, "Being an astronomer, it's easy to get a sense of futility about it all. I am afraid we don't live very long. The species won't live very long, either." He gave a kind of sigh. "When one is a child, one doesn't have a sense of purpose—and that, I suppose, is one of the many extremely good things about being a child. You get a little older and you get a sense of purpose. Young scientists, young people in general, are so terribly *serious*. They think they know where they are going. We become adults when we decide it's not so serious, after all." He paused. "But the sense of wonder never goes away. I guess that's why I like it up here. It's seventy feet closer to the stars."

"Do you ever dream about this telescope?" I asked him.

"Oh, yeah. Sure." He sounded embarrassed. "I dream about it all the time." He said that he experienced recurrent dreams about the Hale Telescope. "The dreams are always set in prime focus, and I don't know whether you would call them dreams or nightmares." He would be sitting on the tractor seat at prime focus and staring down into the mirror through the eyepiece of a camera. "I see a set of red crosshairs with a guide star on them." He said that he might also hear human voices singing the words "et lux aeterna luceat eis"—"May the eternal light shine on them." Listening to the voices, he would punch the paddle, trying to get the crosshairs zeroed on his guide star, trying to see something faint and indiscernible in the mirror, one of those things out there. But he was never able to quite *see* the thing, whatever it was. "What it all means," he said, "I don't know." He turned around and said, "The sky is coming up." The head of astronomical twilight was creeping over the eastern ridge of Palomar Mountain. "We'd better get down," he said.

The lift pulled away and descended. It came to a halt, and Jim Gunn walked onto the floor, moving respectfully beneath the Hale Telescope, covering his flashlight with his hand, because 4-shooter would be taking data until the moment of astronomical dawn.

And dawn had arrived. Maarten Schmidt took command. He stood behind the night assistant and said, "Regretfully, Juan, you may stop the transit whenever you want."

"Starting the pumps," Juan said. A high-pitched whine ramped up. The telescope floated on Flying Horse telescope oil.

Don Schneider stood looking at the video screen. "The sky is holding up pretty well," he said, meaning that he could see galaxies even while twilight took hold. A moment later he said, "We're losing it." The mirror caught the dawn and flared, and the galaxies washed away and vanished from the television screens.

"Take it to zenith," Maarten said.

Juan hit the slew switches.

Don hovered over Juan. "Look at his hands," Don said. "They're trembling with fear."

"Rock steady!" Juan protested, raising his voice over the rumble of the slew. "We are there," Juan said.

"Mirror closed?" asked Maarten.

"The mirror is closed," Juan said.

Don went downstairs to collect the last computer tape of the night. He returned in a moment and placed the tape in the cardboard box.

"Good night, Juan."

"Good night."

"See you tomorrow night."

The astronomers hurried out of the data room and crossed the summit of Palomar Mountain, following a trail that took them down to the Monastery, where they would sleep all day. This was Don's favorite moment: threading among cedars in the cool light.

Juan also liked morning. He enjoyed the feeling that civilization had entrusted the Big Eye to him. He took his time putting the Big Eye to sleep. He closed the dome. He inspected an empty glass jar that had recently held a mass of Oreos. He noted the weather in the Observatory Log: "Fog—partly clear, then clear—light NW wind." He donned his hard hat and stuffed his notebooks into the marinated-jalapeños box and carried the box to his locker. He found a rag in the locker and stepped on the rag. He walked in zigzags underneath the Hale Telescope, holding the rag under his toe, scuffing up the clear puddles of Flying Horse telescope oil that always bled a little from the bearings, a few tablespoons a night.

He was looking forward to watching the morning news with Lily. They would sit at the kitchen table in their house, and she would ask him how things had gone. "Todo fue bien"—all went well. As his day ended, hers began; she taught school on the mountain. He would sleep through the morning, wake up at noon, maybe cut a little firewood, sleep through the afternoon. The Hale Telescope never appeared in his dreams.

He left his time card by the door: he had logged eleven hours. He opened the door. The sky had become a transparent blue. He closed the door behind him and stepped into the parking lot. He saw pink in the east, milky fog in the valleys, and a few late stars fading fast. He liked to look into the blue of morning. This was such a beautiful feeling. That clear color in the air told him that he and all the other night assistants and astronomers on Palomar Mountain had been working hard up there last night. Working hard on the sky. He touched the brim of his hard hat and glanced back at the dome, white and round like an ancient temple, while a thought crossed his mind, by no means for the first time, that he was only paying his respects to a temple of science.

Appendix 1:
Main Characters

(IN ORDER OF APPEARANCE)

Juan Carrasco. Senior night assistant at Palomar Observatory. The man who operates the Hale Telescope for the astronomers. Former barber.

James E. (Jim) Gunn. American astronomer. The only astronomer who has completely distinguished himself as a theorist, as an observer of things in the sky, and as a builder of instruments.

Donald (Don) Schneider. American astronomer. Software wizard. Born on a wheat farm in Nebraska. Became an astronomer because he wasn't any good at farming.

Maarten Schmidt. Dutch astronomer living in the United States. In 1963, he discovered the true nature of quasars, one of the most important discoveries in the history of astronomy.

Eugene (Gene) Shoemaker. American astrogeologist. Discovered the first known asteroid-impact craters on the earth. Expert in asteroids and comets that could hit the earth.

Carolyn Shoemaker. American astronomer, married to Eugene Shoemaker. Top discoverer of comets in history.

Appendix 2:
Glossary

accretion disk. A disk of material rotating in orbit around a massive object such as a planet, a star, or a **black hole**. The rings around Saturn are an accretion disk. A burning accretion disk is thought to form around a black hole in the center of a **quasar**, and this disk is thought to be the source of a quasar's light.

asteroid. Small body of rock or metal in orbit around the sun. A type of **minor planet**.

Balmer series of hydrogen spectral lines. A regular series of sharply defined colors emitted by heated hydrogen gas. When Maarten Schmidt saw these colors in a quasar, in 1963, they told him that quasars are violent, energetic objects located a vast distance from the earth.

black hole. A place where matter has collapsed upon itself and punched a hole in spacetime, out of which no light can escape. Time in a black hole dies or comes to an end.

Cassegrain cage, or **Cass cage.** A room that hangs from the bottom of the Hale Telescope, under the mirror, containing equipment.

Cassegrain focus. A focal point in a **reflecting telescope** located just below the center of the main mirror (i.e., the **primary mirror**) at the bottom of the telescope.

CCD. Charge-coupled device. An electronic silicon light-sensing chip used in place of photographic film in a camera. Extremely sensitive to light.

comet. Small body of ice or icy rock in orbit around the sun. Near the sun, the comet's ice evaporates, creating a tail of gases streaming away from the nucleus of the comet. When comets go extinct (lose their tails) they may be indistinguishable from **asteroids**.

dark matter, or **missing mass.** The main constituent of the universe. No one knows what it is.

dark time. The moonless time of the month.

data room. Small room beside the telescope where astronomers work.

decomposition of light. The making of a **spectrum** of light by passing the light through a prism or other apparatus, thereby smearing the light apart into its constituent colors (wavelengths or energies).

earth-crossing asteroid. An **asteroid** or extinct **comet** traveling on an orbit that intersects the orbit of the earth and is therefore liable to hit the earth.

first light. A technical term from astronomy signifying the moment when starlight is allowed to fall on a new mirror for the first time. Also used metaphorically in this book to mean (1) seeing something for the first time, and (2) the earliest light emitted from objects when the universe was young.

4-shooter. Electronic camera somewhat larger than a refrigerator, designed and built by James Gunn with a team of instrument builders, some of whom are known as the Wizards of the Wastebasket. The camera is installed at the bottom of the Hale Telescope.

galaxy. A vast cloud of stars, gas, dust, and unknown material. Contains up to several trillion stars. Comes in a variety of shapes.

guide star. A bright star used as a reference point for guiding a telescope while the telescope is tracking the sky (which moves overhead, due to the rotation of the earth).

horseshoe bearing. A C-shaped steel bearing, wet with oil, used to support a telescope as it swings around. The horseshoe bearing on the Hale Telescope is the largest bearing of any kind in the world, measuring forty-six feet across.

impact structure. A large eroded or buried impact crater on the earth. Can be more than a hundred miles across.

light. Electromagnetic radiation composed of photons, or packets of energy, which manifests itself as both waves and particles. Includes radio waves, infrared light, visible light, ultraviolet light, X rays, and gamma rays—all are forms of light.

lookback time. The amount of time it takes for light from an object to reach the earth. Looking farther outward into the sky is equivalent to looking backward in time, since the farther one sees outward, the more ancient the image one sees.

Lyman alpha spectral line. A sharply defined peak of colored light coming from hot hydrogen gas such as that found in a **quasar**.

Main Asteroid Belt. A zone of asteroids orbiting the sun between the orbits of Mars and Jupiter.

minor planet. An **asteroid** or a **comet**.

night assistant. Professional telescope operator. Never an astronomer.

Oreo. Staple of the nocturnal diet of Palomar astronomers.

paddle. A remote-control device with buttons on it, which an astronomer holds in her hands and uses to control the fine motions of a telescope.

Palomar Glue (slang). Transparent tape reinforced with threads of fabric. "Palomar Glue is what holds this place [Palomar Observatory] together"—James Gunn.

primary mirror. The main mirror in a **reflecting telescope**, located at the bottom of the telescope tube.

prime focus. A focal point of a **reflecting telescope**, located near the mouth of the telescope.

prime focus cage. A small chamber at the mouth of the Hale Telescope where an observer can sit, looking down onto the main mirror, and where light-gathering instruments may be installed.

Prime Focus Club. A small, mysterious club of unidentified astronomers who claim to have made love with someone while in the prime focus cage of the Hale Telescope.

quasar. A pointlike source of brilliant light of all colors. Quasars are very distant, primeval objects, deep in **lookback time**, near the limit of the visible universe. The word *quasar* comes from "quasi-stellar object." Thought to be a hot, burning core of a galaxy which contains a **black hole** at its center.

redshift. Reddening, or lengthening, of light waves emitted by an object that is receding from earth. Used as a gauge of relative distance to an object in the sky. The higher the redshift, the farther away the object is.

reflecting telescope. A telescope that uses a mirror to gather light.

Schmidt telescope. A telescope design invented by Bernhard Schmidt. It has a corrector plate made of thin transparent glass, placed at the front of the telescope like a circular window. This plate bends the rays of incoming light in such a way as to give the telescope a wide field of view while making it very efficient at gathering light.

seeing, the. A precise measure of turbulence in the atmosphere. The better the seeing, the better a telescope on earth can form a pointlike image of a star.

slew. To move a telescope very quickly across the sky.

spectrum. An image or plot showing light from an object broken into its constituent colors (or wavelengths or energies). See also **decomposition of light**.

structure up there. Thin, transparent clouds veiling the sky. Bad news for astronomers.

Trojan asteroid. A **minor planet** that travels in Jupiter's orbit, named after a hero from the Trojan war. There are two clouds of Trojan asteroids, one cloud traveling sixty degrees *ahead* of Jupiter in Jupiter's orbit (the Greeks), the other traveling sixty degrees *behind* Jupiter in Jupiter's orbit (the Trojans).

Credits

The list is long, but many people deserve mention. First of all, the support, enthusiasm, and sensitive editorial judgment of Morgan Entrekin made this book a reality. I am also indebted to Sallie Gouverneur for her wisdom and for her faith in me.

A number of members of my family provided a most effective combination of moral support and cash for this book when it was in its proto-nebular phase: my parents, Dorothy and Jerome Preston, Jr.; my grandparents, Iva and Jerome Preston, Sr.; my grandmother, Mrs. Richard H. McCann; and my aunt and uncle, Anna McCann Taggart and Robert D. Taggart. For their moral support, thanks to my two brothers: David G. Preston, M.D., who is the only real scientist in this family, and Douglas J. Preston, who is a writer and whose books are an inspiration to me. I also wish to thank my wife, Michelle Parham Preston, for her insightful reading of many parts of my manuscript, not to mention her ability to listen patiently to enough stories about quasars and asteroids to have caused any other wife to consult privately with a lawyer against that day when her husband loses his mind.

For their friendship and encouragement, many thanks to John and Yolanda McPhee, Bonnie Hunter, Bill Howarth, Lewis and Ellen Goble, and Helen and Robert Alexander.

My particular thanks to Prof. John Thorstensen, of Dartmouth College, for his thoughtful reading of many parts of the manuscript for scientific accuracy, and for often helping me to find the right words. Any errors of scientific fact in this book, however, are entirely my own folly.

I owe a debt of scholarship to two historians of science: Spencer Weart, of the American Institute of Physics, and David DeVorkin, of the Smithsonian's National Air and Space Museum. Dr. Weart and Dr. DeVorkin have built up and nurtured fine collections of oral-history interviews with astronomers, physicists, and space scientists, among which Dr. Weart's interview with Maarten Schmidt and Dr. DeVorkin's interview with James Westphal were especially interesting and useful to me.

I owe a debt of another kind to the late Wilbury A. Crockett, a retired teacher of English from Wellesley High School, in Wellesley, Massachusetts. Unfortunately, I was not one of Mr. Crockett's better students, but he somehow managed to instill in us his respect for words. And many thanks to Robert Chambers, of Pomona College, for taking his astronomy class on a field trip to see the Hale Telescope. That was my first meeting with the Big Eye.

Also many thanks to Harry Evans of Random House, for making this new edition of *First Light* possible, and to my remarkable editor at Random House, Sharon DeLano. And thanks to Charlie Conrad at Anchor Books for his enthusiasm and help.

■ ■ ■

A large number of people gave interviews and supplied background material and expertise for this book. Many, many thanks to:

Horace Babcock

William A. Baum

Morley Blouke, Tektronix, Inc.

Eileen Boller

Edward Bowell

Robert Brucato

Bobby Bus

George Carlson

Michael Carr and family

Lily Carrasco

G. Edward Danielson

Edwin W. Dennison

Wilfried Eckstein

Earle Emery

Gene Fair, Fair Optical Co.

Jesse L. Greenstein

Fred Harris

Eleanor F. Helin

Byron Hill

John Hoessel

James R. Janesick

Melvin W. Johnson

Paula Kempchinsky and Patrick Shoemaker

Gillian Knapp

Helen Knudsen

Luz and Alicia Lara

Tod Lauer

David J. Levy

Ernie Lorenz

Mrs. Okla McKee, Historical Archives and Museum of the Catholic Diocese of El Paso, Texas

Brian G. Marsden

Jim Merritt, explorer

Gerry Neugebauer

The night assistants of Palomar Observatory:

Jean Mueller

Jeff Phinney

Skip Staples

J. Beverley Oke

Jeremiah Ostriker

Bohdan Paczýnski

Georg Pauls

Bruce H. Rule

Fred and Linda Salazar

Paul Schechter

James Schombert

Mark Serrurier

Lyman Spitzer

John Strong

David Tennant

Robert Thicksten

Edwin L. Turner

Arthur H. Vaughan

Ludmilla Wightman

The Wizards of the Wastebasket:

    Jovanni Chang

    Richard Lucinio

    Victor Nenow

    J. DeVere Smith

James A. Westphal

Barbara A. Zimmerman

▪ ▪ ▪

Finally, for all of his help, I wish to thank Larry Blakée. When he was twelve years old, he saw the two-hundred-inch mirror being polished in the Caltech optical shop, something he never forgot. When he grew up, he became the first electronics technician for the Hale Telescope—he devoted his working life to that mirror and to all of the things that surround it.

RICHARD PRESTON is the author of *The Hot Zone* (about lethal viruses) and *American Steel* (about the Nucor Corporation's project to build a revolutionary steel mill). He is a regular contributor to *The New Yorker,* and has won numerous awards, including the AAAS-Westinghouse Award and the McDermott Award in the Arts from MIT. *First Light* won the American Institute of Physics award in science writing. An asteroid has been named "Preston" in honor of *First Light*. Preston is a lump of rock the size of lower Manhattan. It is likely to some day collide with Mars or the earth.

A NOTE ON THE TYPE

This book was set in Fairfield, the first type-
face from the hand of the distinguished American artist and engraver Rudolph Ruzicka
(1883–1978). Ruzicka was born in Bohemia
and came to America in 1894. He set up his
own shop, devoted to wood engraving and
printing, in New York in 1913 after a varied
career working as a wood engraver, in photo-
engraving and banknote printing plants, and
as an art director and freelance artist. He
designed and illustrated many books, and was
the creator of a considerable list of individual
prints—wood engravings, line engravings on
copper, and aquatints.